Technocratic Socialism:
The Soviet Union in
the Advanced Industrial Era

Technocratic Socialism

The Soviet Union in the Advanced Industrial Era

Erik P. Hoffmann and Robbin F. Laird

Duke Press Policy Studies

Duke University Press *Durham, 1985*

Printed in the United States of America
on acid free paper

Library of Congress Cataloging in Publication Data
Hoffmann, Erik P., 1939–
Technocratic socialism.
(Duke Press policy studies)
Bibliography: p.
Includes index.
1. Technological innovations—Soviet Union.
2. Technology and state—Soviet Union. 3. Industry and state—Soviet Union. 4. Technology—Social aspects—Soviet Union. 5. Communism and science—Soviet Union.
I. Laird, Robbin F. (Robbin Frederick), 1946– .
II. Title. III. Series.
HC340.T4H64 1985 338'.06 85-4553
ISBN 0-8223-0644-1
ISBN 0-8223-0692-1

TO ANDERS

In Loving Memory of an Extraordinary Boy

Contents

Acknowledgments

Although this volume is the product of many years' work, it was shaped and completed under the most challenging of circumstances. Our efforts have borne fruit only because they were rooted in rich soil. Numerous relatives, friends, colleagues, staff, and students helped us in significant ways. We would like them to know that the depth of our gratitude is reflected in the directness of its expression here and in the pages that follow.

This study completes a trilogy comprised of *The Politics of Economic Modernization in the Soviet Union* (Ithaca, N.Y.: Cornell University Press, 1982) and *"The Scientific-Technological Revolution" and Soviet Foreign Policy* (Elmsford, N.Y.: Pergamon Press, 1982). The present volume includes, with kind permission of the publisher, a substantially revised version of Hoffmann's "Soviet Views of 'The Scientific-Technological Revolution,'" *World Politics* 4 (July 1978), copyright (c) by Princeton University Press.

Introduction: A Soviet Ideology of Advanced Modernization

Western observers have underscored the conservatism of contemporary Soviet politics and the stability of Soviet perspectives, aims, and institutions since the tumultuous regimes of V. I. Lenin, J. V. Stalin, and N. S. Khrushchev. Westerners acknowledge some changes in Soviet policies and policy making under L. I. Brezhnev. But they characterize Brezhnev's initiatives as tactical adjustments enabling Communist party (CPSU) leaders to preserve the "essential" features of the Soviet system in evolving domestic and international conditions. Not surprisingly, such interpretations minimize the actual and potential influence of worldwide scientific, technological, economic, and social developments on the Soviet polity and society.

In contrast, a handful of Western policy analysts and scholars have emphasized the changing nature of official Soviet ideology in the Brezhnev period, especially party leaders' views about what they term "the scientific-technological revolution" (*nauchno-tekhnicheskaia revoliutsiia*) (NTR). Cyril Black affirms that "the most important development in Soviet ideology since 1917 has taken place almost unnoticed by American commentators." He identifies this development as "a new and dynamic appreciation of the role of science and technology as the critical factor in economic and social [progress]." Part of this "significant change in the Soviet world view" is the recognition that physical labor is no longer the chief element in production and that the USSR must look to Western nations for leadership in many areas of research and development. Black concludes that Soviet thinking about the NTR has transformed Marxism-Leninism, not abandoned it.[1]

Assessments such as Black's have been occasionally challenged but more often ignored in the West. A few American researchers and government officials deny the existence of new Soviet perspectives. Some question the practical significance of innovative Soviet ideas and images. Others contend that changing Soviet orientations are important merely as post facto rationalizations or justifications. Still others insist that, Soviet rhetoric notwithstanding, "revolutionary" advances in science and technology have yet to be made in the USSR.

One's position on these issues, be it advocacy or indifference, is integrally related to one's understanding of Soviet politics. Before elaborating our own position, we must note that numerous high-ranking Soviet officials and social theorists have expressed views quite similar to Black's. For example, D. M. Gvishiani, deputy head of the State Committee on Science and Technology and the late Prime Minister A. N. Kosygin's son-in-law, declared in 1977:

> The disclosure of the essence and social role of the scientific and technological revolution is one of the most significant theoretical and political conclusions reached by Marxism in the second half of the 20th century. The further development of this revolution makes the theoretical elaboration of social, economic, ideological, and philosophical questions particularly pressing and also explains the growing interest in these questions on the part of broad circles of the public and on the part of political parties and government organs. *This interest is by no means of an academic nature. The communist and workers' parties are elaborating a theory which is a guide to action and forms the scientific basis of their practical program.* Speeding up scientific and technological progress to the utmost and making use of achievements of the scientific and technological revolution are the main ways of attaining further development of the material and technological basis of socialist society and its growth into the material and technological basis of communism.[2]

Moreover, a senior Soviet theorist emphasizes that scientific revolutions always take place "in scientists' thinking, in their overall mode of perceiving the world they study." In B. M. Kedrov's opinion, the essence of the present-day NTR is the "collapse of faith in the possibility of the development of 'pure' science wholly divorced from the demand of practice [and the] recognition of the necessity of an organic fusion of science, technology, and production." Kedrov maintains that Soviet policy makers and analysts must strive to understand the interconnections among many different branches of the natural, social, and policy sciences, and that they must ascertain the socioeconomic determinants and consequences of scientific discoveries and technological innovations in various types of political systems.[3]

As the Gvishiani and Kedrov statements illustrate, Soviet political officials and researchers are well aware of the recent global breakthroughs in science and technology and of the concomitant need to improve societal guidance in the USSR. Conceptions of the NTR have pervaded the official pronouncements of the highest party and state leaders since the mid-1950s, and these evolving conceptions have influenced Soviet assessments of domestic and international capabilities, policy-making and administrative procedures, and national policy in important fields. Khrushchev, for example, was acutely sensitive to the political implications of modern weapons technology, especially the disastrous international consequences of nuclear warfare and the need to adjust Soviet foreign and domestic policy accordingly (e.g., to seek "peaceful coexistence" with the West). Also, Khrushchev vigorously prodded Soviet natural and social scientists to help resolve practical problems, especially those concerning the performance of the economy. Brezhnev made a greater effort to link theory and practice in the *formulation* of national goals and programs as well as in their implementation. That is, Brezhnev's collective leadership encouraged closer study of the interconnections between ends and means, and Brezhnev (much more than Khrushchev) tailored domestic and foreign policies to the human and material resources available and to the political environments at home and abroad.

Of particular significance were the Brezhnev administration's efforts to understand the role that science and technology are playing and could play in the transition to new forms of "developed socialism" and, eventually, to "communism." Also notable were Soviet attempts to develop a theory of "the scientific management of society" (*nauchnoe upravlenie obshchestvom*) (NUO); to affirm the centrality of human choice and initiative in societal development; and to assess the positive and negative consequences of scientific and technological changes in socialist, capitalist, and Third World countries. Key premises of Brezhnev's report to the 26th Party Congress in 1981 were that scientific and technological advances are critical to economic advances and that "the interrelationship between economic progress and the social, political, and spiritual progress of society is becoming ever closer."[4]

Soviet theorists examine important linkages between theoretical and practical issues, and they raise key questions about the interaction among political-administrative, socioeconomic, and scientific-technological changes. Soviet investigators analyze the reciprocal relationships among people, science, and technology; the preconditions and stimuli of scientific, technological, economic, and social progress in different political systems; and the features that distinguish the contemporary NTR from industrial revolutions of the past as well as from possible "new" scientific and technological revolutions in the future. Moreover, Soviet analysts pay considerable attention to the evolving modes and components of scientific, social, and managerial thought; the changing nature of labor; the use of new technologies in economic decision making and implementation; the NTR's impact on the social structures of different polities; and the effects of the NTR on individual capabilities and personalities and on humankind's relationship to the physical environment.[5]

The Brezhnev leadership encouraged innovative thinking by proclaiming in 1967 that the USSR's "socialist" phase had ended and that a "mature" or "developed socialist society" had been launched. Top Soviet officials and analysts stressed that theory and practice must actively influence one another. The major text on dialectical and historical materialism affirmed that Marxism-Leninism "is a creative, revolutionary doctrine, a doctrine that is constantly being enriched and tested by historical practice. It is opposed to any kind of dogmatism and constantly develops on the basis of generalization of the experience recorded in world history and the achievements of the natural and social sciences."[6]

We believe that it is fruitful to analyze the intellectual content of authoritative Soviet views on advanced industrialized societies; the social and ideological consequences of these perspectives; and the reciprocal effects of ideas, policies, and administrative behavior during the Brezhnev period and thereafter. We hope to elucidate for Western readers a cognitive and problem-solving orientation that is strongly encouraged by CPSU leaders, is a major evolving component of Marxism-Leninism, is characterized by a diversity of assessments, and is a means of shaping and justifying the values, attitudes, and beliefs of the Soviet political and educational elites. The functions and especially the content of Soviet ideology shifted considerably under Brezhnev, and we will examine these shifts and some of their origins and outcomes.

This book completes a trilogy on contemporary Soviet views of political, administrative, social, economic, scientific, and technological progress. In our first volume we focused on the rivalry between Soviet economic reformers and conservatives, emphasizing the international dimensions of their different strategies of modernization.[7] In our second volume we focused on Soviet foreign policy, emphasizing the changing and competing concerns of party leaders and their advisors about the NTR's impact on international politics.[8] In the present volume we focus on Soviet domestic policies, on decision making and implementation, and on parameters of consensus among leading party officials and social theorists.[9]

We scrutinize the concept of "developed" or "mature socialism," emphasizing the common ideological framework within which cognitive reorientation and conceptual innovation are taking place.[10] Calling for diverse but circumspect and pragmatic new ideas, Brezhnev almost always supported a centrist interpretation. Although he modified his outlook from time to time, it consistently influenced or molded Soviet policies and policy-making procedures. For example, fresh ideas about the NTR had a major impact on the 24th Party Congress program, which initiated the period of East-West detente in the early 1970s. Also, traditional orientations are often of considerable policy relevance because Soviet ideologues hold much more powerful political positions than do Western social theorists. Hence, we have reconstructed the evolving world view of leading party officials and analysts and have assessed its influence on contemporary Soviet politics.[11]

We concentrate on four major topics: the NTR, developed socialism, scientific management, and the mass media. A chapter is devoted to each of these topics and to some of their interrelationships—in Soviet thinking and in ours. We begin with primarily theoretical and perceptual issues and progress to primarily practical and behavioral concerns. The final chapter compares Soviet and Western ideas about modern industrialized societies and summarizes our views about the linkages between theory and practice. Stressing the importance of politics and eschewing technological determinism, we contend that Soviet perspectives changed much more than Soviet institutions during the Brezhnev years and afterward. We emphasize the increasingly pragmatic nature and functions of official Soviet ideology, the shifting roles of the party and state, and the expansion of elite participation in policy making and of mass participation in administration. We conclude that the Soviet Union is becoming a new type of authoritarian order in the advanced industrial era, and we term that order "technocratic socialism."

1
"The Scientific-Technological Revolution" and Political and Social Change

The purpose of this chapter is to analyze the Soviet concept of the NTR, especially the questions it raises and its success in resolving such questions. The NTR's significance in Marxism-Leninism is clearly reflected in a Soviet theorist's contention that "the development of mankind's productive forces has gone through three milestone epochs, which are known as the neolithic, the industrial, and the scientific and technical revolutions."[1] Although Soviet commentators have scrutinized the interplay between scientific-technological, socioeconomic, and political-administrative developments, they have yet to produce a comprehensive theory of the NTR. Two prominent Soviet scholars, introducing a collection of essays on the NTR, observe: "We hope that an examination of the different [Soviet] approaches to the problem of the scientific-technological revolution will enable us to develop a uniform conception of it in the future."[2] Hence, we must examine the areas of agreement among Soviet analysts, the issues in dispute, and the trends in this policy-relevant discussion.

Soviet political leaders and theorists stress that significant changes in "the material forces of production" have taken place since World War II and are continuing rapidly worldwide. Foremost among these developments are the remarkable scientific discoveries of new sources of energy and materials and technological innovations in countless fields. Also deemed crucial are changes in the characteristics and activities of human beings—their values, aspirations, capabilities, knowledge, work, and cultural pursuits. Like Karl Marx, Soviet commentators study the effects of specific transformations in productive forces on "the social relations of production" (authority, property, or "class" relations) and, in turn, the effects of production relations on the "superstructure" (the political culture and institutional arrangements) of various societies (socialist, capitalist, and Third World). Moreover, Soviet analysts examine the reciprocal influences of the superstructure on production relations and of the production relations on productive forces of various social orders. Although Soviet observers consider these influences to be less important than the impact of the economic "base" on the political, legal, and military superstructure, they ask pertinent questions about science, technology, and production in highly industrialized nations (e.g., are economic activities part of the superstructure or of the base?). Furthermore, socialism and capitalism are believed to be radically different socioeconomic systems—the former grounded on "relations of cooperation and mutual assistance," the latter on "class antagonism and class struggle."[3]

A general Marxist-Leninist orientation is shared by all Soviet theorists of the NTR, but there is ample debate about important specifics, linkages, and research and development priorities. For example, what are the actual relationships among the scientific, technological, environmental, and human productive forces of contemporary Soviet society? What are the most desirable interconnections among the evolving productive forces, production relations, and superstructure of a one-party polity in a rapidly changing international system? The complexity of such issues—to say nothing of their various policy and ideological implications—suggests that CPSU leaders have been wise to encourage some creative theorizing and empirical study of the NTR and its socioeconomic consequences. Indeed, Soviet theorists are using Marxist-Leninist analysis to deepen their understanding of what developed socialism and communism can and should be, and of how present-day scientific-technological and socioeconomic forces must be harnessed in order to bring existing conditions closer to shifting conceptions of an ideal polity and society.

All Politburos after Stalin's have sanctioned a certain degree of ideological innovation, cognitive readjustment, and conceptual search. Officials and theorists have not always used key concepts in similar ways. They have not repetitively asked the same questions, and they have not arrived at identical answers. Indeed, a major collective study identifies and briefly contrasts diverse Soviet viewpoints on "the essence, content, basic trends, historical role, causes and consequences of the contemporary scientific-technological revolution."[4] Although Soviet analysts do not overtly challenge traditional values and precepts, they attempt to rethink or "concretize" basic concepts and ideas under present-day conditions. As Julian Cooper emphasizes, "One cannot yet speak of a single, generally accepted Soviet theory of the scientific and technical revolution and its social consequences."[5] Hence, Soviet leaders and theorists are posing, studying, and disputing questions whose answers are not presumed to be known beforehand.

A few Soviet commentators have suggested that Soviet research on the NTR consists primarily of description rather than analysis, generalizations, and theoretical conclusions; that there has been insufficient empirical investigation of key aspects of the NTR; and that there should be a moratorium on future studies until greater agreement can be reached about the nature of the NTR and the appropriate methodologies for studying it. Other Soviet commentators have retorted that such views advocate a particular interpretation of and a methodological approach to the NTR. Also, they argue that data gathering is linked inevitably to one's theoretical orientation and that a well-developed general theory should guide the collection and interpretation of data about the NTR.[6] Although such arguments may constitute special pleading, they are responses to the post-Stalin leaderships' encouragement of limited "methodological pluralism" and of competition among alternative ideas about the NTR.

The Nature of the NTR

What have Soviet social theorists accomplished to date? First, there has been a prolonged and in many ways fruitful discussion of the fundamental characteristics or "essence" of the contemporary NTR. Frequently based on serious study of the history of various technologies, these works demonstrate a growing understanding of the origins and significance of recent scientific discoveries and technological innovations, of the importance of the social contexts in which they are applied, and of the methodological and practical problems of analyzing the cultural, economic, and political ramifications of scientific-technological progress in socialist and capitalist countries.

An authoritative Soviet definition of the NTR is offered by P. N. Fedoseev, a full CPSU Central Committee member since 1961 and vice-president of the Academy of Sciences of the USSR:

> The scientific and technological revolution is basically the radical, qualitative reorganization of the productive forces as a result of the transformation of science into a key factor in the development of social production. Increasingly eliminating manual labor by utilizing the forces of nature in technology, and replacing man's direct participation in the production process by the functioning of his materialized knowledge, the scientific and technological revolution radically changes the entire structure and components of the productive forces, the conditions, nature, and content of labor. While embodying the growing

> integration of science, technology, and production, the scientific and technological revolution at the same time influences all aspects of life in present-day society, including industrial management, education, everyday life, culture, the psychology of people, the relationship between nature and society.[7]

More specifically, a major Soviet study identifies six central components of the NTR and offers a lengthy categorization of phenomena that express or reflect these components. The six fundamental or defining characteristics are:

1. The merging of the scientific revolution with the technical revolution, predominantly influenced by developments in science (manifestations are the "scientification" of technique and the "technification" of science);
2. The transformation of science into a "direct productive force," which makes possible the scientification of production (reflected in changes in the tools of labor, the "materialization" of the objects of labor, sources of energy, technological applications of science in production, human capabilities, and management and organization of production);
3. The organic unification of the elements of the production process into a single automated system whose actions are subordinated to general principles of management and self-management;
4. A qualitative change in the technological basis of production, signifying changes in man-machine relations that are stimulated by increased use of materialized knowledge and by enhanced human capabilities to manage and control production processes;
5. The formation of a new type of worker who has mastered scientific principles of production and can ensure that the functioning of production and its future development will be based on the achievements of science and technique;
6. A major shift from "extensive" to "intensive" development of production, such as broader and deeper application of scientific and technical advances, more efficient use of natural resources, and sharply increased labor productivity.[8]

Such conceptualizations leave open for theoretical development the likely and desirable effects of the new productive forces on the production relations and superstructures of different societies and on the "dialectical" relationships among these three major clusters of variables. In general,

Soviet theorists view the NTR as a necessary but not a sufficient means of achieving higher stages of developed socialism and, eventually, communism. The NTR is to be "mastered," and to do so the nature of the opportunities and obstacles it presents must be better understood. For example, A. G. Egorov, a candidate or full member of the CPSU Central Committee since 1966 and director of its Institute of Marxism-Leninism, contends that both the natural and social sciences are increasingly becoming "direct productive forces." In a mature socialist society, Egorov argues, science influences the productive forces through comprehensive economic and social planning as well as through innovative technological processes and products. The social sciences help to develop the productive forces by enhancing citizens' theoretical knowledge, technical know-how, and labor productivity.[9]

Soviet theorists have made concerted efforts to clarify key concepts such as "science," "technology," and "revolution" and to identify the NTR's features and manifestations and the interrelationships among them. Some Soviet commentators subsume numerous empirical referents under the concept of the NTR and thereby reduce its analytical fruitfulness. Others fail to distinguish between new technical advances and the larger social and cognitive systems in which they must operate. And a few observers, such as Fedoseev in an early edition of his magnum opus on the contemporary era, rarely refer to the NTR.[10] But one can observe in countless Soviet pronouncements the creative and competitive redefinition and reapplication of concepts. The questions to be asked, the methods of answering them, and the answers themselves are subjects of ongoing dispute. Different intellectual approaches to the NTR often reflect political disagreements among party, state, and production officials. Hence, terminological discussion about "science," "technology," and "revolution" are clear signs of ideological adaptation and of vigorous debate within the parameters of officially approved general ideas about the NTR. Significantly, most Soviet analysts criticize the substance of each other's interpretations rather than their "deviance" from an a priori norm. And, despite many vicissitudes and lacunae, increasingly sophisticated and comprehensive theories of the NTR emerged in the Brezhnev years.

Scholars from Moscow's Institute of the History of the Natural Sciences and Technique (*Institut Istorii Estestvoznaniia i Tekhniki*) (IIEiT) played a key role in this development, and they made an important distinction between "technical," "production," and "social revolutions."[11] Not surprisingly, Soviet theorists consider the social revolution launched in October 1917 to have been the essential precondition for the creation of new production relations and for the "progressive" use of previous scientific discoveries and technological innovations. A technical revolution is a major set of advances in "technique" (i.e., tools, production methods, and labor- and energy-saving devices) that transforms the productive forces. A production revolution is a cluster of significant breakthroughs in "technology" (i.e., the social organization of labor, production, and management as well as the know-how and skills needed to apply knowledge from the physical and social sciences to feasible goals and tasks). Such a revolution is to help create new production relations but cannot maintain them without a thoroughgoing social revolution. According to the IIEiT theorists, "History shows that a technical revolution can sometimes precede a social revolution, but a production revolution can begin only after a social and technical revolution."[12]

Analysis of this kind reflects an understanding that technological innovations cannot be easily grafted onto existing labor, production, and management methods, which must change considerably in order to generate and assimilate new machines, materials, and inventions. Soviet theorists maintain that a technical revolution can be achieved only if adjustments are made in the larger social systems of which the modern techniques are a part. For example, a computer will surely not produce "revolutionary" changes in administrative power relationships and decision-making practices unless it is accompanied by changes in the content and flow of information and in bureaucratic attitudes toward secrecy and security.

Soviet commentators insist that new styles of scientific thought and technological modes of production are having and will continue to have a tremendous effect on economic and social progress. N. V. Markov asserts that "technique is produced by technology, not vice versa. Technology is predominant in this relationship."[13] Such a view supplements that of the IIEiT scholars because it is highly sensitive to the influences of production relations on productive forces, especially of human/machine relationships and the division of labor on the uses of new techniques. Also, some Soviet theorists are placing greater emphasis on the social and political contexts in which scientific technological innovations are generated or developed. Markov implies that techniques of capitalist and nonsocialist origins are having a much greater impact on technologies in the USSR today than they will in the future. He is among the small but growing number of Soviet analysts who have begun to discuss the "limits" of the contemporary NTR, its probable completion in "less than 100 years," and the coming of a "new" NTR that will be more distinctively socialist and will have even more positive consequences than the present-day NTR.[14]

Such perspectives reflect increasing Soviet concern about the environmental problems posed by the NTR and about the need to improve relationships between humankind and nature in socialist societies. Cooper observes that more and more Soviet theorists regard the contemporary NTR as "a transitional phenomenon, a final upsurge of an historically limited 'industrial' mode of production, to be superseded by an 'ecological' mode of production brought about by the reorientation, or 'ecologization,' of all the sciences and technology."[15] Cooper concluded in 1981 that "there is a coexistence of diverse Soviet theories with ecologized variants increasingly displacing the more traditional conceptions."[16]

Almost all Soviet theorists are keenly aware of the larger cultural contexts of scientific and technological progress. Serious scholars and propagandists alike attempt to explain why the NTR speeds the transition to communism in socialist societies but exacerbates the "antagonistic contradictions" and hastens the collapse of developed capitalist nations. The basic Soviet presupposition is that collective ownership of the means of production, national planning, and broad public participation in the management of society make possible an equitable distribution of human and material resources. Because capitalist governments promote the interests of powerful "monopolies," research and development are concentrated in the military and consumer goods spheres, and scientific and technological advances fail to serve the interests of the majority of the people.[17] V. G. Afanas'ev, a CPSU Central Committee member since 1976 and editor-in-chief of *Pravda*, asserts: "The bourgeoisie can manage efficiently individual enterprises or large corporations or even whole industries, but it is incapable of managing the entire national economy and society as a whole. The reason for this is the domination of private property, the laws of anarchy and competition which preclude conscious management on a scale of the whole society."[18]

According to Soviet analysts, scientific and technological developments are value-free but are embedded in value-laden social and political systems that utilize them for purposes such as corporate profit or the common good. Hence, the applications of technologies are determined by the prevailing social and political values, and the consequences of scientific, technical, and production innovations vary considerably in different societies.[19]

While emphasizing the need for "radical" change in certain cognitive processes and production relations, NTR theorists affirm that once socialism has been established, the dominant political, economic, and social values and structures are relatively impervious to retrogressive change. Significantly, they argue that the NTR is bringing another production revolution to the USSR. The transfer of ever more physical and mental work from people to machines and to automated management and control systems is to create "fundamental social transformations in society."[20] Soviet commentators conclude that socialism gradually evolves into communism because collectivist norms and relationships stimulate both scientific-technological and socioeconomic progress. "It is possible to construct communism only by fully utilizing the opportunities the NTR presents to society—[and only by] *mastering the revolution itself and learning to manage it*."[21]

Politics, Society, and the NTR

Soviet theorists underscore the political influences on the uses of science and technology, the need to plan and coordinate social and economic development, and the diverse socioeconomic consequences of the NTR. Fedoseev affirms that "the Marxist approach to the NTR centers not on the problems it presents to society (which would separate the NTR from society, place it over society as an external unrelated force), but rather on what tasks society, with its vital requirements of continued economic, social, and spiritual growth, sets science and technology, thus revolutionizing their development."[22] V. Cherkovets declares that "the improvement of management is not a technical but a profoundly social process, and its aims and the course it takes are, in the final analysis, determined by the social system of production. At the same time, the creation of these forms is *an active quest*, a planned approach to the improvement of management."[23] V. V. Kosolapov adds that "the current scientific and technological revolution in itself does not solve any pressing problems facing society . . . and has no bearing on the solution of the question of the ownership of the means of production and *the distribution of material and cultural values*, including the achievements of science and engineering. These questions can only be tackled by a social revolution."[24]

The most distinctive aspects of Soviet thinking are the presumed manageability of the NTR and the accompanying assurances that party leaders can control the social and economic outcomes of global scientific and technological changes. Soviet theorists consider socioeconomic progress in the USSR and allied nations to be an "objective necessity," which can be accelerated by "subjective" factors and by the application of scientific-technological advances to "progressive" national goals. According to Soviet analysts, the fruits of the NTR are value-free, and, regardless of the political system in which technologies are developed, they will be put to good use under socialism and harmful use under capitalism. Hence, CPSU spokespersons emphasize the political context in which modern science and technology are applied, and their arguments and assumptions have both practical and intellectual consequences.

The NTR concept facilitates Soviet discussion of obstacles to progress in the USSR. Officials and specialists often refer or allude to the "new and complicated problems" of socialist societies as well as to the "many serious problems" of capitalist states. Soviet commentators acknowledge that the NTR produces negative physical consequences in both capitalist and socialist countries (e.g., exhaustion of fossil fuels, pollution of the air, water, and soil, and ecological imbalances caused by industrial uses of oxygen and fresh water).[25] A leading Soviet geographer asserts:

> In the epoch of the modern NTR and of the rapid rate of industrialization and urbanization, society's impact on the environment has attained unprecedented intensity and proportions, with a steady snowballing tendency. There have appeared local and regional *scarcities* of various natural resources, and irreversible disruptions and destructive spontaneous processes have appeared in the environment on a regional and *global* scale. Among these disruptions and processes dangerous and harmful changes worsening the conditions of mankind's life have appeared and have been rapidly growing. In consequence, a real threat has emerged in the environment of a disruption of the natural material and energy balances and ecological conditions of life, which tends to complicate the further existence and activity of human society.[26]

I. Gerasimov cites important and mounting problems such as the excessive consumption of raw materials, the "industrial voids" caused by strip mining and the dumping of industrial waste, the effluence of heat into the atmosphere and bodies of water, the depletion of minerals in the soil, overcrowding and noise pollution in the cities, and insufficient natural recreation areas. Gerasimov maintains that "in none of the industrial *capitalist* countries have the effects of the growing impact on the environment been overcome." Expressing cautious optimism about socialist systems, he observes that "the Marxist-Leninist ideology enjoins consideration for nature, man's habitat and its resources."[27]

Also, there have been occasional Soviet warnings about "technocratic and bureaucratic" behavior among the scientific, technical, and economic elites of socialist countries. According to Iu. E. Volkov, "In the course of the NTR, technocratic and bureaucratic tendencies can arise, [but] it is not fatally inevitable that they become stabilized. Their neutralization, however, should not be considered an easy or almost automatic matter. . . ."[28] Volkov is concerned about the possibly harmful beliefs, attitudes, and values that the NTR can engender in the Soviet state and society. Like Markov and others, he focuses on the processes rather than on the end products of management, production, and labor. Hence, some Soviet theorists reject the view that the NTR can have no unintended or undesired effects on "the distribution of material and cultural values" (cf. the passage cited in note 24).

Furthermore, Soviet critiques of capitalism provide clues to Soviet apprehensions about the possible negative consequences of the NTR in socialist countries. Especially because of Brezhnev's emphasis on the transfer of whole *systems* of technology from West to East, G. N. Volkov's statement may be more revealing than intended:

> Capitalist society has not only created a highly developed machine system but complicated social mechanisms as well, which "grind" the individual as efficiently as machine production. As a consequence, a man does not control his machine tool—the machine tool controls him, that is, it dictates to him the nature and order of machine-like actions and determines the speed and rhythm of the labor process. Moreover, he is the object of bureaucratic management at all levels.[29]

Soviet commentators also contend that automation and labor specialization increase alienation and "technological unemployment" in capitalist states. They acknowledge that extensive retraining and redistribution of human power and a difficult period of adjustment of perhaps several decades will accompany the widespread introduction of automated production systems in the USSR. But, according to Soviet spokespersons, these modern systems and new kinds of specialized labor will eventually enhance work productivity, job satisfaction, leisure time, individual self-fulfillment, and social harmony.[30]

How does one explain the very different social and political consequences of automation in different societies? Most Soviet officials stress the dominant role of production relations. But conservative theorists hint that imported technologies may be producing more similar cross-national effects than party leaders anticipated. These analysts emphasize that vigilant selection, installation, and monitoring of innovative production and organizational technologies are needed to preserve the fundamental characteristics of the Soviet polity, economy, and society. Such commentators stress that modern political-administrative and socioeconomic mechanisms must direct or control, rather than be unduly influenced by, the NTR.

At bottom, Soviet theorists make an important claim—namely, that scientific, technological, economic, and social progress are not inevitably or quickly achieved, and that only a unified society led by the CPSU and under "public control" can "make full use of the positive aspects of the scientific and technical revolution for the benefit of man and neutralize its possible negative consequences."[31] For the serious theorist, the linkages between scientific-technological and socioeconomic development are important components of the concept of "mature socialism" that he or she is trying to elucidate and disseminate. For the ideologue, these linkages are axiomatic. Consequently, Soviet observers insist that political and organizational factors greatly influence the NTR's development in all societies and that the CPSU can and must guide the use of technical advances, native and imported.

But Soviet theorists pay little attention to the values that may be inherent in or closely associated with the successful functioning of complex machines and production processes. They also deemphasize the diverse political, economic, and social contexts in which new techniques and technologies are generated. In other words, Soviet spokespersons stress that the social and organizational contexts in which the NTR's achievements are *used*, not those in which they are *developed*, affect people's values, interpersonal relations, and material well-being.[32] Also, Volkov declares that certain technological innovations "indirectly" increase labor productivity through changes in production relations, whereas other techniques "directly" influence the content and organization of labor "relatively independent of socioeconomic relations; one and the same assembly line calls for highly specialized, mechanical operations, no matter whether it is installed in a Detroit plant or a plant in Sverdlovsk."[33]

Compare Volkov's assertion with one attributed to an American businessman in the USSR and cited by a Soviet writer in 1946 as an example of a "very widespread" method of ideological subversion. "We like your manganese. Your manganese has one incredibly remarkable feature. It doesn't know it is socialist. It is just as ready to go into the furnace in Pittsburgh as in Stalingrad. You like our machine tools in just the same way. They don't know they are capitalist. They are just as ready to cut metal in Kharkov as in Detroit." The Soviet observer characterized the American's statement as an effort "to subject the social relations of people to the relations of things."[34] Particularly because of the xenophobic cultural policies of Stalin and A. A. Zhdanov, 1947 to 1953 was probably the only period in Soviet history when the transfer of technology from West to East was officially deemed a harmful form of cultural penetration. Lenin, shortly after the October 1917 revolution, advocated a memorable formula for the construction of socialism. "To scoop up with both hands the best from abroad: Soviet power + the Prussian railway system plus American technique and the organization of trusts + American education, etc., etc. + + = Σ = socialism."[35]

Briefly stated, Soviet analysts have traditionally affirmed that technique is value-free, technology has some values inherent in or associated with it, and socioeconomic production relations are the dominant sources or allocators of values. They argue that there are no technologically determined social relations. In a developed socialist society, collectivist norms are to govern all human and human/machine relationships, and the technological and natural environments are to enhance the citizenry's material and spiritual well-being.

Soviet officials and specialists are acutely aware that the NTR is developing most rapidly in the United States, Western Europe, and Japan. They acknowledge, sometimes grudgingly, that Western technology and management practices can help to realize "the advantages of the socialist economic system." But Soviet observers confidently predict that modern products and processes will strengthen the fundamental characteristics of the present Soviet polity, economy, and society. Soviet analysts affirm that imported and indigenous technological innovations will serve only those purposes that the highest party and state leaders deem to be in the best interests of the Soviet people. Accurate or not, this contention is politically and intellectually significant because it emphasizes the "neutrality" of techniques and their capacity to produce different socioeconomic outcomes in capitalist, socialist, and developing countries.

Although Soviet commentators are quick to emphasize the diverse socioeconomic consequences of technological progress in various polities, it is apparently imprudent to suggest that dichotomies such as direct/indirect, objective/subjective, and form/content are themselves highly subjective and value-laden. These dichotomies are reminiscent of the controversial Western positivist distinction between "facts" and "values." Ironically, Soviet theorists may be employing a Western mode of thought to deny the existence of undue Western influences. To be sure, many Soviet analysts have stated or assumed that scientific and technological advances can be easily transferred among states with different social systems. But they insist that the NTR's end uses and effects are decisively shaped by political values and institutions, which vary greatly from country to country. Kedrov, S. R. Mikulinskii, and I. T. Frolov conclude: "The scientific-technological revolution does not develop in a 'social vacuum' and, in turn, exerts a powerful influence on the development of the existing social relations."[36]

Theory and Practice

What are the linkages between Soviet ideas and experience, and how do party leaders and social theorists think theory and practice should influence one another? Responding to such basic questions, Soviet analysts have "creatively" developed Marxism-Leninism and hence have posed further questions about scientific-technological, socioeconomic, and political-administrative developments.

An important case in point is the "historic" task Brezhnev set forth at the 24th Party Congress in 1971: "*to combine organically the achievements of the scientific-technological revolution with the advantages of the socialist economic system* and to develop more broadly our own, intrinsically socialist, forms of combining science with production."[37] This maxim has been frequently reiterated but is challenging to interpret. How does one "combine organically" particular discoveries and innovations with existing modes of production? How does the organic fusion of science and technology alter the structure and functioning of the economy? If, as seems likely, these initiatives are means of pursuing larger ends, the question immediately arises, "organic combination for what?" The familiar Soviet answer is, "to achieve higher forms of 'developed socialism.'" But this response raises additional questions about the nature of developed socialism. What should a more fully developed socialist society strive for, given the values of the present-day bureaucratic elites? What can mature socialism be, or what is it likely to be, given domestic and international constraints and trends? How might forecasting and "prognostics" help to elucidate national goals and enhance the Soviet leadership's capability to fulfill them?

Brezhnev's Politburo sought to integrate the NTR with the essential elements of the existing polity and economy and thereby to increase the population's quality of life. To these ends, theorists and officials were urged to develop fresh ideas about desirable and possible syntheses between modern technology and fundamental characteristics of Soviet society; to reassess key concepts such as progress, leadership, and management; and to clarify the multifaceted implications of newly formulated maxims, especially Brezhnev's major declaration at the 24th Party Congress discussed above. Sociologists and economists were to provide leaders with the expertise needed to establish priorities to the year 2000 as well as to implement and evaluate programs more effectively and efficiently.

The highest CPSU officials expected members of the scientific-technical and bureaucratic elites not only to study and forecast the linkages between technological advances and the transition to developed socialism, but also *to contribute to an understanding of what the goal itself was*—that is, to help conceptualize what "developed socialism" could and should be and how best to attain it under increasingly complex, dynamic, and uncertain domestic and international conditions. Soviet commentators responded to this challenge in countless books and articles, and they offered an even wider variety of ideas through private official and unofficial channels. Party, state, and production executives participated in a circumscribed but ongoing reexamination of short-, medium-, and long-term national goals and of different strategies to pursue alternative futures. Cautious or unimaginative analysts applied general principles to specific cases and to changing internal and external contexts. Other observers merely engaged in popularization. But many of the questions raised were theoretically significant, and most of the answers—whether or not they were intellectually innovative—were policy-relevant or performed diverse social and ideological functions.

For example, the following authoritative Soviet assertions concern the national policy-making process and the relationship between knowledge and power:

> The action of the objective laws of history is inseparable from those *goals* that people (classes and social groups) set for themselves and strive for, and for the *means* they employ to pursue these goals. . . .
>
> The present scientific-technological revolution makes possible further improvements *in the methods of formulating goals of social activity*, primarily by creating the conditions for increasing the information potential of society. *Comprehensive* investigation of production and social processes, and the processing of large amounts of information, are absolutely necessary *before* actions are taken. *The very act of decision making regarding social goals is itself a research process*, which is effective to the extent that it conforms to the demands of a scientific approach. . . .
>
> The difficulties involved in social management consist primarily in the fact that one has to solve not only given tasks and to secure the implementation of the goals set, but also *continuously to work out new tasks and set new goals in the course of social development*. It is becoming increasingly evident that man, as the subject of management, cannot be replaced by the machine as a matter of principle. This does not

> exclude, at the same time, the enormous importance of utilizing cybernetic ideas and principles for the organization of management.[38]

Such Soviet statements are unusual because they acknowledge that certain public policies must be reshaped or rejected in response to the vast quantities of new information generated by scientific and technological advances and their socioeconomic outcomes. In other words, some Soviet analysts recognize that national goals can no longer be established by the intuitive application of traditional precepts and that if one brings more and better data about the relationship between ends and means into decision making, *the goals themselves may have to be reformulated.*[39] Social relations and human motivations are thought to be undergoing a gradual but considerable transformation, in response not to technical imperatives but to the planned use of scientific-technological breakthroughs and to the reconceptualization of socioeconomic aims. "*The dialectical interaction between goals and means is manifesting itself especially clearly at the present time. . . . The spurious alternative of the 'technical' and 'value' content of social goals must be totally rejected. . . .* Socialism demonstrates the possibility and feasibility of establishing an organically combined program of social and scientific-technological progress."[40]

But most Soviet analysts stress that pertinent and timely information must be used to carry out traditional policies and that established decision-making and implementation procedures must be adapted to unanticipated policy outcomes and new opportunities. Soviet commentators insist that political leaders need tremendous amounts of information about the functioning of the polity and society in order to manage, utilize, and spur the development of "objective social laws" and "progressive trends." This approach focuses on technically competent administration of centrally determined priorities and "the politics of details," rather than on innovative elite and mass participation in policy making and "the politics of principles." Hence, competing Soviet ideas about the uses of information now reflect or influence leading Soviet officials' perspectives on basic policy and policy-making questions.

For example, the Brezhnev administration sharply distinguished between the purportedly value-free and value-laden components of work, production, and management. These distinctions justified the expanded importation of advanced technology from the West, explained the growing international division of labor, and denied the possibility of subversive influences on Soviet citizens. Some Western observers conclude that CPSU leaders were thereby rationalizing to themselves and to the population the shortcomings of the Soviet science-technology-production cycle vis-à-vis the accomplishments of capitalist nations. Less charitably, other Westerners suggest that the Soviet rhetoric about the NTR was primarily intended to obscure the fact that no NTR has taken place or is likely to take place in the USSR, even with the aid of Western technology, credits, and trade. And still other Western critics contend that Soviet theories legitimize institutional relationships that *impede* scientific and technological progress.

The Soviet leadership surely does not wish to retard scientific-technical and economic development. The Brezhnev administration encouraged creative theorizing about the NTR while seeking to preserve existing industrial and agricultural institutions and practices. Ideas about the NTR and NUO may have misguided Soviet officials and sustained their hopes that sophisticated technical advances (e.g., computers) could strengthen current institutional and social structures (e.g., centralized economic planning and management). But if Soviet conceptualizations of the NTR had such effects, they were not merely post facto justifications; they were part of the motivational ideology of some, perhaps many, party and state officials.

Indeed, Soviet leaders' perspectives on the NTR seem to have become an increasingly salient and unmanipulable part of their cognitive orientations. Brezhnev and his colleagues did not merely try to rationalize or deny the existence of problems; they attempted to understand, anticipate, and resolve dilemmas and to take advantage of opportunities. The collective leadership strove to formulate and implement science and technology policies that furthered widely accepted political goals (e.g., modernization and stability) and responded to competing priorities, interests, and constraints. Also, top Soviet officials encouraged the selective adoption of Western know-how and equipment, and they exhorted bureaucrats and workers to overcome administrative obstacles to economic growth and productivity. To be sure, the Brezhnev leadership made only minor changes in the formal and informal rewards and sanctions that undergirded and often undermined the practical application of scientific and technological advances. But the Politburo and Secretariat seriously endeavored to fuse the achievements of Western science and technology with the social and economic structures of the USSR. The fact that leading CPSU cadres pursued this goal cautiously and in ways that most Western observers deem ineffective does not mean that Soviet ideas about scientific-technological and socioeconomic change were post facto rationalizations. On the contrary, the Soviet leaders' evolving perspectives on the NTR appear to have shaped or reinforced diverse preferences concerning public policies and policy implementation.

Reformers such as Kosygin wanted to take firm action to reduce organizational "irrationalities," to "optimize" economic decision-making and administrative procedures, and to adjust selected party and state institutions and relationships. Centrists such as Brezhnev agreed that Western methods of production, management, and planning were essential to Soviet economic development and that capitalist technical advances could be effectively adapted without compromising socialist values. But Brezhnev believed that changes in administrative behavior could take place within the *existing* political and economic structures and with less export-oriented production than Kosygin wanted. Conservatives such as M. A. Suslov were warier of organizational innovations, imported technology, global markets, and the purported benefits of Western economic and managerial approaches. Suslov was especially apprehensive about the social and cultural consequences of the growing interdependence between capitalist and socialist states.[41]

Yet the social functions of ideas can change considerably over time. This is what happened to the democratic and populist aspects of communist ideology, which were a viable policy alternative immediately before and after the October 1917 revolution and which later helped to create and legitimize Stalin's personal dictatorship.[42] Likewise, the purposes and effects of NTR theories have changed moderately since the introduction of the concept in the mid-1950s, and future changes are quite possible. For example, in 1957 Khrushchev used the term "NTR" to justify the elimination of most national ministries, the establishment of regional economic councils, and the expansion of the CPSU's role in regional economic planning and management. Rhetoric about the NTR was also employed in the 1960s in conjunction with the formation of the State Committee on Science and Technology and with the unrealistic economic and social goals of the CPSU's new third program.[43]

In the late 1960s the greatly increased official commentary about the NTR was part of a campaign to formulate more effective national policies and to mobilize bureaucratic support for the major decisions of 1969–1971 (e.g., the effort to promote economic growth and productivity by substantially increasing East-West trade, foreign credits, industrial cooperation agreements, and the importation of advanced technology). In the mid- and late 1970s, the NTR concept became a key element in the evaluation of the consequences, many of them undesired and unanticipated, of these policies. Throughout the 1970s ideologists popularized ideas about the NTR in order to enhance public support for the party's guiding role in the political system and for specific party directives.

Assessments of the NTR influenced the Soviet leadership in a much more conservative way in the late 1970s and early 1980s than a decade earlier, and the aging CPSU leaders employed NTR theories chiefly to legitimize previous decisions. Soviet perspectives on modern science and technology could play a more conservative or reformist role under M. S. Gorbachev and the new generation of top- and middle-level party officials who, succeeding Brezhnev, Iu. V. Andropov, and K. U. Chernenko, must confront mounting socioeconomic problems at home and greater political and economic competition from abroad. Although the NTR concept has been used less frequently in the 1980s, it is slated for inclusion in the CPSU's fourth official program, which has been in preparation since 1981. Shortly after becoming general secretary, Chernenko declared, "The idea of the fusion of two revolutions—the scientific-technological and the social—should be given fitting reflection in the CPSU Program."[44]

Briefly stated, diverse Soviet views of the NTR can justify *and* shape actions. Such views can rationalize the preservation of the status quo and motivate a leader to avoid change, *and* they can rationalize and motivate various kinds and degrees of innovation. The term "NTR" underscores that worldwide scientific and technological advances make possible, even necessary, "radical" or "revolutionary" transformations in many areas of Soviet life. Afanas'ev, an influential centrist, maintains that "the scientific-technological revolution has a profound effect on the process of formulating and making decisions and places exceedingly great demands on decisions."[45] Therefore, if the current or future Soviet leadership wishes to initiate far-reaching policy or institutional innovations, many of the concepts and ideas needed to mobilize CPSU members and to explain to others why certain changes are desirable, feasible, prudent, or inevitable are already at hand.

Since Stalin's death the broad outlines of NTR theories have been so well developed and official support for these theories so strong that future leaders who might want to dispose of the NTR orientation will be faced with a difficult task. The NTR concept could, of course, be transferred from "realistic" ("practical," "operative") ideology to "grand" (i.e., "pure," "fundamental") ideology.[46] But the content and history of this perspective generate pressure for ongoing ideological innovation of both the "guide to action" and the post facto rationalization varieties. This pressure is intensified by the NTR's rapid development in capitalist countries, by the Soviet leadership's and citizenry's growing awareness of Western scientific and technological advances, and by the increasing permeability of the Soviet economy and society to external influences (e.g., global economic disturbances and cultural innovations disseminated by modern telecommunications).

In contrast to the fate of the utopian communist "goal culture" of the early Bolsheviks, the trend toward more justificatory uses of NTR theories is not irreversible. The balance between the motivational and the rationalizing ideological functions of the NTR concept, which links an "ideology of ends" and an "ideology of means" much more closely than did its Bolshevik predecessors, depends on many variables. The complexity of contemporary scientific-technological and socioeconomic problems, opportunities, and interrelationships, and the distasteful but distinct prospect of a widening manufactured and consumer goods gap between the USSR and the major Western industrial nations, were among the key factors that induced the Brezhnev administration to alter Khrushchev's policy-making and organizational procedures. Brezhnev's collective leadership seriously weighed advice and data from "responsible" members of all the major Soviet bureaucracies, and it included representatives of these bureaucracies in the Politburo. Although all Politburos choose their own theorists and specialists, they need multifaceted expertise to grapple with the NTR.

To be sure, most Soviet officials and analysts tell political leaders what the leaders want to hear. But, especially under Khrushchev, there was a diversity of political orientations at the highest levels of the party. Also, Brezhnev, Andropov, and Chernenko enabled more and more Soviet bureaucratic and educational elites to participate in comprehensive social and economic planning as well as in policy implementation and evaluation. Social scientists, employing survey techniques and working closely with party and state agencies (including the secret police), became increasingly involved in assessing the scientific-technological and socioeconomic consequences of policies and in molding and monitoring public opinion on a wide variety of issues. General Secretary Gorbachev might initiate and carry out major changes in Soviet politics, with researchers and other specialists playing a predominantly propagandistic *or* prescriptive role.

The expanded interaction between Soviet politicians and experts is likely to have a growing influence on the thinking, judgment, selection, and even the power of national political leaders. Members of the Politburo and Secretariat are aware that scientific and technical factors are of great importance in most public policies. They are also aware of the difficulties of "managing" persistent and potentially serious socioeconomic and scientific-technological problems in rapidly changing international and moderately changing domestic environments. Likewise, NTR and NUO theorists view politics from a national rather than a local or departmental perspective. They urge that only the most basic decisions be made at the center and that technical expertise and relevant social information be brought to bear on policy making and implementation at all stages and levels and on virtually all issues.

Important organizational and economic sources of power other than scientific-technical information persist. Also, the inevitable disagreements among experts enhance the autonomy of politicians vis-à-vis experts. But, as the worldwide NTR develops, national and regional CPSU officials will need new scientific, technical, and managerial skills and a deeper understanding of current economic realities (e.g., market forces). Specialists have begun to make greater technological, administrative, and political contributions to decisions in many fields, and they will probably continue to do so through greater consultation, cooptation, and direct representation in party organizations and soviets. At the same time, the CPSU leaders' perspectives on appropriate national policies and policy-making procedures have been shifting. These perspectives will continue to change, especially as top party and state cadres establish new criteria and standards to evaluate "scientific," "rational," and "effective" political-administrative decisions and new methods to cope with "problem situations" (Afanas'ev's term).

Far greater policy and institutional changes will be necessary in the future if there is to be a thoroughgoing NTR in the Soviet Union—that is, if the nation is to contribute more fully to and to benefit from what is still a predominantly capitalist NTR. The USSR already has been able to copy many contemporary scientific and technological breakthroughs, enhance its role in the international economy, become a global military power, and promote its conception of social progress at home if not abroad. These are basic aims of the Soviet leadership that NTR theories reflect and sustain.

An even more "secularized ethos"[47] in the ideological thinking and calculations of Soviet elites is likely to develop in the USSR. To the extent that party leaders continue to stress "comprehensive" social and economic planning, theories of the NTR will play a major role in identifying trends and future conditions. Forecasting will clarify the value trade-offs implicit in policy options and will help to reformulate ideological principles in anticipation of policy changes. Ideas about the NTR also may perform a major justificatory function in legitimizing the shifting relationship between the party and state. Furthermore, the highly instrumental NUO theories may have a considerable impact on the operative ideologies of Soviet leaders and production officials, especially if innovative approaches, such as systems and cost-benefit analysis, produce tangible economic results.

If socioeconomic problems mount and if "objective" or perceived pressures for change increase, the Politburo and the Secretariat might considerably broaden and deepen elite participation in policy making, expand permissible policy alternatives, and mobilize bureaucratic support for experimental ideas that would alter scientific-technological and socioeconomic planning and management. In the event of a major crisis, unorthodox ideas about the NTR and NUO could play a decisive role in changing elite orientations and national policy. The leadership's strong desire for political stability and its predisposition to resort to established methods will continue to provide powerful support for the status quo. At the same time, CPSU leaders will consider it pragmatic to think in new ways about "stability," "interdependence," "democratic centralism," "collective leadership," and "party guidance." They may also acknowledge that traditional organizational practices and technology are impeding high-priority objectives and eroding the legitimacy of the political system.

If scientific, technological, social, and economic failures persist, and if competing Soviet leaders come to believe that these problems are insurmountable without substantial changes in institutional relationships and in domestic and foreign policy, ideas about the NTR might have even greater influence. In bottleneck and crisis situations, NTR theories will probably mold rather than justify policies and policy-making procedures. After all, economic problems at home and the NTR's achievements abroad helped to adjust the criteria and standards by which the Brezhnev administration judged scientific-technological, socioeconomic, and (to a much lesser extent) political-administrative "success" and "failure."

It is both a fact and an article of faith that the NTR is an agent of portentous change. Top Soviet political leaders and social theorists have been exploring ways to cope with forces they do not fully understand, but whose importance they clearly sense. Together they are working out methods of adjusting traditional values and procedures to international scientific and technological developments that are beginning to influence, and to be influenced by, production advances in the USSR. Soviet spokespersons are increasingly aware that bureaucratic predispositions and practices—and possibly certain political values and institutions—must be creatively adapted to the NTR and vice versa.

Conceptualizations of the NTR are part of the Soviet leadership's frame of reference for defining situations, identifying problems, evaluating options, adjusting policies on how to make policy, advocating political-administrative programs, and assessing the consequences of previous decisions. But, because top CPSU officials have not fully supported a particular theory of the NTR, different interpretations abound and the NTR concept remains elastic. Party and state leaders have underscored *various* linkages between scientific-technological and socioeconomic change, and their ideas have supported *various* aims, assumptions, and actions. Policy groups have developed vested interests in specific interpretations of the NTR, and these diverse perspectives have been incorporated into realistic or practical ideology. Many competing interests and perspectives notwithstanding, there is a crucially important common denominator in the Soviet bureaucratic elites' thinking about the NTR. They have agreed that national political goals must remain paramount, that scientific and technological progress must be viewed in instrumental terms, and that new technologies must be centrally controlled or approved.

Policy and Policy-Making Implications of the NTR

Soviet theories of the NTR reassess the fundamental characteristics of the contemporary era and redefine the nature of the changing circumstances in which the party leadership must act. Soviet analyses of social, technical, and production revolutions are of considerable political significance. Especially important are Soviet elites' views about the interconnections among different kinds of revolutionary change, the senses in which they are revolutionary, the extent to which "nonantagonistic contradictions" intensify pressures for change, and the degree to which these changes can be controlled in the short and long run. Also, do Western methods of thinking and production impart or contain undesirable "capitalist" values? How successfully have Western management practices been adapted in the USSR? And how successful *could* they be if institutional reforms were carried out?

Top party and state leaders' responses to such issues are quite likely to influence perceptions of options and choices among them. Soviet officials contend that socioeconomic and scientific-technological progress in complex industrialized societies can result only from centralized planning and decision making, which in turn demand clear thinking about medium- and short-range political goals and about how to progress from existing to desired conditions. Brezhnev repeatedly called for improved social and economic planning and forecasting, which was to be based on better statistics and extrapolation of current trends and on a clearer exposition of basic goals and ends-means relationships.[48] Also, Afanas'ev argued that "the social system's functioning is anticipatory, and the controller must constantly have his eye on the future."[49] And Gvishiani has consistently stressed the "inseparable" links among forecasting, planning, and management.[50] Hence, Soviet thinking about the future will probably help to shape the nature and pace of current developments. The traditional Soviet ideological commitment to planning, the perceived "spontaneity" of the NTR, the political challenge of controlling the NTR's socioeconomic consequences, and the mounting scarcities in human and material resources may induce Brezhnev's successors to conceptualize national goals and to assess national capabilities in increasingly long-range terms.

A particularly important question is whether and how the development of Marxist-Leninist theories of the NTR will influence the selection of party policies and the basic institutional relationships and policy-making procedures of the Soviet polity. Scientific and technological advances and the ideas of the NTR theorists are least likely to change the political leaders' cognitive orientations and basic values, more likely to change selected policy-making methods and attitudes, and most likely to change specific policies and beliefs. The nature of political-administrative discourse in the Soviet Union is evolving, but a major shift in the values of the bureaucratic elites—whether produced by some kind of scientific and technological "imperative" or by political pressure from social and natural scientists—does not seem imminent or probable. Only a substantial change in some of the basic characteristics of the Soviet political system (e.g., the proscription of associational interest groups and the *nomenklatura* or privileged job lists and functionaries) is likely to break established patterns of administrative behavior.

Bureaucratic coalitions dispute the meaning of the NTR because different ideas have different policy implications and practical consequences. In his speech to the 25th Party Congress, Brezhnev suggested that recent scientific and technological developments had changed the perspectives of high-level Soviet officials more than those of their subordinates. "The revolution in science and technique requires fundamental changes in the style and methods of economic work, a determined struggle against sluggishness and red tape; *it requires true respect for science* and the ability and desire to take advice from and reckon with science."[51] Apparently, top party leaders were unable or unwilling to make the prevailing bureaucratic culture—especially the career and communication patterns—accord with their evolving and divergent views about modernization. In any case, top Soviet officials recognized that scientific and technological advances could have an enormous impact on national policies and on policy making and administration. And the party leadership, without changing the traditional patronage and incentive systems, continuously exhorted state and production executives to utilize advanced technologies.

Although Western scholars know little about the politics of Soviet theorizing about the NTR and NUO, these pronouncements are not only part of the "ideological struggle" with the West but are a key component and arena of internal political competition over power and policy. Even in the factory or on the farm, young managers with freshly learned techniques claim to be "agents" of the NTR, thereby imposing their preferences on other executives and workers and downgrading the informal administrative skills and technical know-how of experienced personnel. The undesirable consequences of these developments (e.g., animosity, mistrust, reduced productivity) are a theme in contemporary Soviet fiction[52] and are receiving closer attention in Soviet sociological research.

The editors of *Voprosy filosofii*, a major scholarly journal, have stressed "the great practical and political significance" of the NTR literature, especially "the practical value" of formulating a "descriptive, explanatory, and prognostic" theory of the interrelationships between the NTR and Soviet society.[53] Also, as Gvishiani pointedly asserts, the purpose of research on the NTR is "not only to publish monographs and articles, but to work out concrete suggestions, recommendations, and methods that planning organs, ministries, and departments can use."[54] Is the pragmatic benefit of Soviet theorizing more persuasive justification of current policies and policy-making methods? Or is the benefit evident in new policies and procedures that promote changing official values or updated applications of traditional values? On these salient questions and on numerous specifics, Soviet analysts are reticent. To be sure, they stress the growing importance of political leadership in the present and future stages of developed socialism, and they assert that the beliefs and attitudes of political leaders and bureaucrats are changing. But the possibly evolving values of Soviet elites are treated publicly with the utmost circumspection. Only in Soviet conceptualizations of more "developed" forms of socialism are the present official goals, the gaps between theory and practice, and the political organization of society sometimes *implicitly* called into question. For example, Chernenko observed in 1984: "The formation of the classless socialist society will be an important stage on the path to complete social homogeneity and will, no doubt, entail *substantial changes in all superstructures*."[55]

Western observers emphasize the need to study the interaction between society and technology as well as the structures and processes linking technical innovations ("the part") to the larger social context in which they are used ("the whole").[56] Soviet theorists strongly agree in principle but, like their Western counterparts, often do not heed their own advice. Much Soviet forecasting is founded on questionable assumptions and inadequate consideration of existing conditions in the Soviet Union and elsewhere. Many Soviet commentators base their analyses largely on the officially desired uses of technology and on ideologically prescribed values, tenets, and trends. To be sure, Soviet researchers conducted more empirically grounded and middle-range theoretical studies of the party and state bureaucracies in the Brezhnev years. But analysis of the relationships among organizational technology, design, and performance is not one of the strengths of these investigations. Too often Soviet social scientists merely assume that a homogeneous socialist society, party, or state *is* generating scientific-technological innovations and *is* utilizing these opportunities exclusively for the public good. Problems of departmentalism, localism, coordination, communication barriers, role conflict, and center-periphery relations deserve—and are beginning to receive—more serious attention in Soviet analysis of the NTR and NUO.

"Images of the future"[57] have been important elements in the orientations of Stalin's successors and are central components of Soviet theories of the NTR. In contrast to the wishful thinking of the Khrushchev years, Brezhnev's collective leadership focused on feasible aims and on the trends, not the end results, of scientific-technological and socioeconomic change. That is, the choices of goals and means were more closely intertwined, and the conceptualizations of developed socialist society were more consistently linked to the *transition* to mature forms of socialism.

The Brezhnev Politburo directed social researchers to deepen the leadership's understanding of desirable courses of action and to offer practical recommendations (when requested) on how to implement them. Fedoseev declared, "The elaboration of a national development program is impossible without the active participation of the social scientists. Whether it deals with a country's energy or agricultural potential, or demography or culture, social science in its economic, social, and ideological components plays an essential and sometimes dominant role."[58] Also, shortly before becoming general secretary, Chernenko affirmed:

> Our social science is called upon to study questions of paramount theoretical and practical significance, such as the nature and types of contradictions that are characteristic of Soviet society's contemporary stage of development and the objective and subjective factors that cause them, and not to try to dismiss all the accompanying difficulties and negative phenomena as "survivals of the past" in people's con-

> sciousness. Multifaceted theoretical analysis of such problems helps the party to understand more deeply the economic, social, or other causes of the contradictions that arise and to surmount them in a timely and effective manner.[59]

Many Western observers consider the Brezhnev administration to have been inflexible and lethargic, especially in its final years. However, Westerners overemphasize the remarkable continuity in institutional structures and top-level personnel, and they underemphasize the incremental changes in the perspectives of and the relationships among various officials and specialists. The cumulative impact of even minor cognitive and interpersonal changes can be significant. Indeed, the Brezhnev leadership's domestic and foreign policies and policy-making procedures were considerably influenced by its evolving perceptions, expectations, and calculations about the complex phenomenon it termed the NTR. A prime example, as we have noted, is the coordinated set of domestic and foreign policy objectives presented at the 24th Party Congress in 1971.

CPSU leaders have yet to devise successful means of nurturing the NTR or of enhancing ample creative rather than duplicative capabilities. Soviet theorists stress, however, that production relations decisively shape a society's initiatives and reactions and "mediate" the development, applications, and effects of science and technology in different countries.[60] At bottom, Soviet theories of the NTR reiterate that scientific-technological and socioeconomic progress are closely linked and that major advances in both spheres can be achieved through production revolutions in socialist systems and through social revolutions in capitalist systems. To the extent that such ideas are supported by investigations of actual conditions in various contemporary societies and can help to explain the reciprocal influences between domestic and international change, the emerging theories of the NTR will make a major contribution to Marxism-Leninism.

We conclude that the Brezhnev administration, especially in the early and mid-1970s, encouraged serious theorizing and research on contemporary scientific-technological, socioeconomic, and political-administrative change. Soviet commentators addressed broad intellectual and practical issues concerning technology and politics. The content of official Soviet ideology evolved, and the more pragmatic new ideology shaped as well as justified political-administrative choices. In other words, descriptions and analyses of the NTR are innovative and ideological in two ways: they serve as "a guide to action" and as a rationalization of acts motivated by other factors. Leading party officials and theorists are unlikely to agree on a single, systematic interpretation of the NTR. Even a consensus on the key issues and methods of resolving them is problematical. But Soviet Marxism-Leninism will surely be refined and possibly be revised as scientific and technological breakthroughs continue to transform much of the world. Evolving perspectives on progress are core elements of the theory and practice of developed socialism and are the subject to which we now turn.

2
"Developed Socialism" and Progress

We will now develop two themes that were briefly introduced in the previous chapter. We noted that the concept of the NTR was integrally related to the concept of developed socialism in the Brezhnev years and that the Soviet belief in progress focuses on the "dialectical" relationship between the productive forces of the NTR and the production relations of socialist society. The first section of this chapter describes and analyzes Soviet perspectives on mature socialism. The second section investigates the tensions and ambiguities inherent in a concept of progress that fuses the "universal" NTR with traditional Soviet values and institutions.[1]

The Soviet Concept of Developed Socialism

For nearly twenty years after the October 1917 revolution, party leaders characterized Soviet Russia as a society "constructing socialism." In 1936 Stalin proclaimed that "socialism" had been established in the USSR because private ownership of factories and farms had been virtually eliminated. The nationalization of industry and the collectivization of agriculture were the core elements of Soviet socialism in its initial phase. In 1967, however, Brezhnev declared that maturing human and material productive forces had created a "developed socialist society." According to Fedoseev, "The conjunction of the NTR's achievements and socialist social relations creates the conditions and premises for the establishment of the unity of man and nature on a new and higher technical, scientific, and social basis."[2] G. N. Volkov affirms that modern science has become a force for socialist transformation because it can be "oriented to man and to changing and improving his biological and social nature and the environment in which he lives."[3] A. I. Berg asserts: "The fact that under socialism science and its technological applications serve the interests of society as a whole, with the scientific approach dominant in all spheres of socialist society, elevates to an unprecedented degree the social significance of science and the responsibility of scientists."[4] V. I. Kas'ianenko adds:

> Developed socialism is not an independent or self-sufficient phase but only a new stage in the maturing of socialist society. The construction of developed socialism occurs after the complete transition from capitalism to socialism. Developed socialism speeds our transition to communism by immeasurably accelerating the tempo of social progress, ensuring a high level of the productive forces and production relations, mobilizing the powerful potential of the NTR, and, finally, creating the conditions for the successful communist upbringing of members of the new society.[5]

Using a stage-of-development approach, Soviet theorists contend that evolutionary change is needed to perfect the already legitimized political organization of society. Developed socialism broadens and deepens socialist structures and property relations under new conditions. Existing institutions adapt to "revolutionary" scientific and technological advances. Soviet analysts call for the incremental modernization of key institutional relationships and for the thoroughgoing modernization of many administrative practices. These changes are to improve the performance of the political system's fundamental and secondary elements and to maintain political stability.

Soviet conservatives and reformers view the NTR as a beneficial force. Conservatives hope the NTR will strengthen the centralized features of the Soviet system; reformers hope the NTR will modify those features. Conservatives underscore the need for *continuity* in the social and organizational structures of socialism and developed socialism. They look to the NTR to preserve the essence of the existing Soviet polity and society in an increasingly competitive world that rewards *self-sufficiency*. Reformers underscore the need for *change* in the social and organizational structures of socialism and developed socialism. They look to the NTR to alter the traditional Soviet political-administrative procedures and social structure in an increasingly competitive world—a world that rewards *interdependence*. Conflict between these two orientations is minimized only when the substantive and procedural changes associated with the NTR are left unspecified or when general appeals are made to use the NTR's achievements to strengthen the USSR's fundamental characteristics in the long transition from socialism to communism.

Individual and Societal Needs. Soviet theorists agree that science molds objective reality through technological processes. Only under mature socialism, Egorov contends, "can the *advantages* of a new social structure and the *possibilities* of socialist progress *fully unfold* on the basis of a powerful economic and scientific-technological potential and an affirmation of socialist principles in all spheres of life."[6] Furthermore, the overcoming of "nonantagonistic contradictions" is central to the Soviet conception of progress. These contradictions include "(1) those that arise from class and social distinctions; (2) contradictions between individual interests; (3) between group interests; (4) between immediate and long-term interests; (5) between the national and international interests of the working class; (6) between objectively conditioned needs and their refraction in the consciousness and behavior of people."[7]

Another nonantagonistic contradiction is between the economic and cultural dimensions of development. Science-based high technology increasingly influences cultural needs, and the cultural prerequisites of technological breakthroughs and their timely use are rising steadily. For example, managers and technicians require higher levels of education to spur economic growth and productivity. More scientists, engineers, and skilled workers and more effective use of their talents are essential to the NTR. But technological advances are much slower than the evolving possibilities revealed by science, and economic performance is much weaker than the evolving possibilities revealed by technology. Hence, a contradiction between the availability and application of knowledge must be overcome in order to speed socioeconomic development.

Soviet analysts maintain that mature socialist societies consist of two friendly classes—the industrial workers and the collective farmers—together with a burgeoning social stratum, the intelligentsia. The latter is the segment of the working class that is engaged in predominantly mental labor. R. I. Kosolapov, a full or candidate CPSU Central Committee member since 1976 and editor-in-chief of the party's major theoretical journal, *Kommunist*, observes that the intelligentsia, "once a small stratum, is growing into a substantial and highly heterogeneous social group, which, in the USSR, is numerically greater now than the class of *kolkhoz* peasants due to the growing demand for competent mental labor."[8] The increasingly large and diversified scientific-technical intelligentsia (not to be confused with the cultural intelligentsia) has contributed greatly to economic modernization. Scientists, engineers, and managers are a leading innovative force and are a vital component of the bureaucratic and educational elites.

The Soviet scientific-technical intelligentsia is to enhance "the progressive unity" of society, which the NTR and the intelligentsia's expanding role in production make possible. Soviet theorists contend that under capitalism the intelligentsia is alienated or coopted by state monopolistic institutions. Because of the cleavages among the classes of capitalist society, the intelligentsia's creative talents can only serve private interests. Under mature socialism, however, party and state organs can ensure that "the socialist intelligentsia . . . does not seek to 'monopolize' all intellectual activity with the aim of turning it into a factor of power, but, on the contrary, spares no effort to make intellectual wealth accessible to all working people, so that by obtaining up-to-date knowledge and acquiring education and culture they will be better equipped to exercise democratic state power."[9]

Soviet commentators affirm that the CPSU can foster social unity by directing and monitoring the activities of the scientific-technical intelligentsia and by helping to meet the increasingly similar needs of the intelligentsia, proletariat, and peasantry. For example, NTR theorists maintain that the rapidly growing stratum of service workers generates nonantagonistic contradictions in a developed socialist society, but that the West's service personnel are joining the working class in all-out opposition to state-monopoly capitalism. Also, pure and applied research are thought to be furthering collective goals under developed socialism but furthering corporate and personal goals under advanced capitalism.

Markov and other Soviet analysts argue that the NTR substantially alters the nature and functions of work and reduces dissimilarities between intellectual and manual activities. Industrial laborers acquire sophisticated skills, and enterprises are more and more automated. Agrarian workers become industrialized, even urbanized, as mechanized farms increasingly resemble large factories. Trends such as the merging of mental and physical labor, the automation of production processes, and the easing of manual work are thought to be furthering a "many-sided, systematic, and consistent 'scientization' of labor, which is based on data from numerous sciences ranging from the socioeconomic to the anthropological and biomedical."[10]

The relationship between the scientization of production and the development of culture is critical for Soviet theorists. They emphasize that dialectical and historical materialism elucidate the ultimate metaphysical reality and provide the only correct world view. This orientation purportedly gives socialists a distinct advantage over capitalists, whose "outmoded" religious beliefs impede scientific rationalism. In the USSR scientific and technological progress may be blocked by antiquated modes of action but not by an "antagonistic" metaphysical or socioeconomic system. The Soviet people are to strive for "more effective use of all the factors of economic growth, more effective and rational use of the means and objects of labor and their constant improvement, more rational and effective use of the forces of nature and of the environment. . . ."[11] Hence, instrumental approaches to the physical and social sciences are a core value of developed socialism.

Mature socialism's emphasis on scientific rationalism provides fertile ground for expanding the scope of technical control over nature, including human nature. N. Nikolaev calls for a "substantial enrichment of the organizational principle of a truly scientific approach to activity, to social phenomena, and to the scientific optimism that spurs the individual to self-perfection. This principle is an integral component of the Marxist-Leninist worldview."[12] Volkov declares: "The socialization of science has a nonantagonistic and universal character because it takes the form of the 'scientization of the entire society' or, in other words, the restructuring of society and the creation of a new mix of social relationships that conforms to the demands of science and to the laws upon which it is based."[13]

According to Soviet analysts, progress in the material base of society can lead to social progress only through developed socialist production relations. The primary component of socialist production relations is the public ownership of the means of production. Collectively owned land and machines make possible the guidance of scientific-technological and socio-economic development for the common good. Elements of planning have been introduced in advanced capitalist societies, but, because of the market economy and monopolistic corporations, planning is segmented and can only serve the interests of the few. Under developed socialism, centralized management of industry and agriculture is to ensure that all citizens benefit from the NTR.

A second component of developed socialist production relations is the commitment to equality.[14] During the construction of socialism, the party-state tried to eradicate the feudal and capitalist values and institutions of the tsarist period. During the 1930s huge financial and professional rewards for performance and loyalty produced a highly stratified society, especially in the burgeoning urban centers and new collective farms. Stalin stressed material incentives before and after World War II; Khrushchev stressed moral incentives in the 1950s; and Brezhnev and his successors continuously struck a balance between the two. Smaller wage differentials and a higher standard of living are now thought to constitute progress.[15] V. Kelle declares that egalitarianism rests "on the growth of production and culture and on perfecting the forms of social organization. This signifies the achievement of social equality, not on a low, but on a high level of production and the comprehensive development of all members of society."[16] In short, contemporary Soviet theorists view greater socioeconomic equality and material abundance as compatible ends and as means of enhancing productivity and creativity as well as collective and individual well-being.

Inequality is to persist for a long time in developed socialist society. Kas'ianenko cites numerous Soviet studies to support his claim that "classlessness" cannot be attained in the USSR in the foreseeable future. Rather, there is to be a gradual equalization and harmonization of the nation's nonantagonistic social strata. By strengthening socialist production relations, the NTR fosters economic, social, and political equality. Kas'ianenko concludes: "Developed socialism provides for the highest possible level of cooperation among classes and social groups and creates the conditions for their further drawing together, for a continual breaking down of barriers between them, which is the basic manifestation of the formation of social unity."[17]

A third element of developed socialist production relations is the further expansion of the planned society. Planning promotes but does not ensure the clarification of values, the pursuit of national and rational goals, and the optimization of ends-means relationships. While introducing scientific and technological innovations that greatly enhance the capabilities of the economy, contemporary political leaders and specialists are to preserve the industrial base. Even more important, social and economic planning are to be integrated in order to coordinate socioeconomic development more effectively and efficiently. According to A. Bykov, "The NTR calls for an essentially new attitude to the function of management, which becomes comprehensive so as to include social, economic, scientific-technical, ecological and other aspects."[18]

Soviet officials and theorists assert that "the political system of socialism, like any rationally organized system, presupposes leadership organs that provide organization, unity, and singleness of purpose for the entire system."[19] During the NTR the bureaucrat, researcher, and worker are to further economic growth and productivity as well as the harmonious development of the social and physical environments. The rational aspects of personality are to dominate over the spontaneous and emotive aspects. Increased control over self and nature in the pursuit of collectivist ideals is deemed a basic cultural and psychological need of "the new Soviet man." Hence, the fourth dimension of mature socialist production relations is the individual citizen's desire and ability to fulfill the socioeconomic goals set by CPSU leaders.

In sum, the "developed" component of developed socialism refers chiefly to the human and material productive forces of modern society, whereas the "socialist" component refers chiefly to advanced production relations. Mature socialism legitimizes guidance of socioeconomic development by the institutions established in the period of socialist construction. The party and state are not to "wither away" in the foreseeable future. References to the party's dissolution are nonexistent in Soviet analyses of developed socialism, and references to the state's dissolution are infrequent.[20] Rather, the developed socialist polity is characterized by a high degree of organizational complexity and by the evolving but increasingly important role of the CPSU in societal guidance—notably, in its relationships to the state bureaucracies and mass organizations. The party's capacity to manage these relationships is claimed to be a new source of its legitimacy and power vis-à-vis other institutions as the NTR unfolds.

Furthermore, mature socialism focuses on the means of achieving feasible ends. Although the ultimate goal of socioeconomic development is communism, Brezhnev and his successors have paid scant attention to this distant ideal. Instead, they have concentrated on the "realistic" conceptualization and "businesslike" guidance of a developed socialist society. Kas'ianenko contends that the political system of developed socialism can fulfill its potential by strengthening the ties among the party, state, and mass organizations and by improving the specific forms and content of their work.[21] That is, the institutional mechanisms and procedures of a modern socialist order are essential components of progress toward communism. According to an authoritative Soviet collective work, "*Developed socialism is at the same time developing socialism, the functioning mechanism of which is inseparable from the mechanism of its development*."[22] Hence, the maturation of socialism refers to the CPSU's growing capability to lead a cohesive society in a rapidly changing world.

"Socialist Democracy": The Party and the State. Soviet theorists emphasize that the volume, interconnections, and dynamism of political decisions are increasing the importance of specialist and mass participation in policy making and administration.[23] Identifying the main challenges confronting the Soviet polity in the era of the NTR, G. Kh. Shakhnazarov, a Central Committee department official and president of the Soviet Association of Political Sciences, affirms:

> The number one task is to further democracy—to secure broader participation of people in decision making, broader equality, and broader rights and freedoms of individuals. Task number two is to improve the state machinery and to optimize the administrative process. The two tasks are interrelated. The people have a vital stake in

> efficient administration, be it of the country as a whole or of individual sections of its economy. The most effective means of improving administration is to involve in it the broadest possible mass of people.[24]

Soviet analysts contend that to accomplish these goals a society needs a single political party that is highly respected by all classes and strata and that has the authority to induce or compel all organizations to serve the public interest. Tsarist Russian and Soviet political traditions have always rejected the Western idea that power, if concentrated, will be abused. The contemporary official Soviet view may be summarized as follows: in a capitalist nation, under a government that furthers only the material well-being of the upper and middle classes, organizational power corrupts and mass political activities are a sham; in the Soviet Union, under a one-party system that furthers the material and sociopsychological well-being of all citizens, organizational power and public participation in administration must be enhanced.

The role of the party is carefully elaborated in the USSR Constitution of 1977:

> The Communist Party of the Soviet Union is the leading and guiding force of Soviet society, the nucleus of its political system and of state and public organizations. The CPSU exists for the people and serves the people.
>
> Armed with the Marxist-Leninist teaching, the Communist Party determines general prospects for the development of society and the lines of the USSR's domestic and foreign policy, directs the great creative activity of the Soviet people, and gives their struggle for the victory of communism a planned, scientifically substantiated nature.
>
> All Party organizations operate within the framework of the USSR Constitution.[25]

According to Soviet theorists, the party embodied the will of a maturing working class under the tsarist autocracy (1898–1917), under a "dictatorship of the proletariat" (1917–1961), and under a "state of the whole people" (1961 to the present). Stalin's "cult of personality" suppressed the party for a prolonged period (1934–1953), but Khrushchev reaffirmed its primacy and Brezhnev regularized its relations with state, mass, and production organizations. The party is to maintain its "leading role" under developed socialism. Industrialization is largely completed, and conflict among hostile classes is no longer deemed a problem. The CPSU is to formulate policies that express the general will of the Soviet people and induce all organizations to help carry out these policies. NTR theorists contend that the party is the only Soviet institution capable of performing such functions in a "scientific" manner, and this claim is frequently used to explain or justify the CPSU's predominance in the polity.

The party is to assume ever greater responsibilities as the "leader," "organizer," "decisive coordinator," "regulator," "synthesizer," "adjuster," "teacher," "mobilizer," and "energizer" of an increasingly well-educated and politically sophisticated population. In Kas'ianenko's view, "The growing role of the CPSU is caused by the tremendous scale and complexity of the tasks of social management, the enhanced role and influence of the working class, the increased significance of scientifically substantiated policies and economic and social forecasting, . . . the growing organizational role of the entire political system, and the deepening of the scientific-technological revolution."[26]

The party's primary task is to formulate just and realistic policies, to spur their implementation, and to monitor their outcomes. As Soviet society becomes more structurally differentiated and its people more knowledgeable, the integrating functions of the party become more important. Top party bodies are to transform into viable policy alternatives the diverse "interests" of the proletariat, peasantry, and intelligentsia and of the subgroups within and among these three "chief elements" in the social-class structure of a mature socialist nation. Shakhnazarov warns that the nonantagonistic contradictions generated by competing strata in the USSR "may on occasion become extremely sharp unless prompt and sensible steps are taken to resolve them."[27]

Furthermore, the CPSU is to elicit information from all organizations and social strata in order to help its leaders determine which goals and policies are in the national interest and how best to pursue them. The party apparatus is to control data, recommendations, and political-administrative options, their expression and interpretation, and the channels through which they are communicated to decision-makers. But some Soviet commentators perceive the need for the political leadership to be more responsive to different interests and to make greater use of scientific, technical, economic, and social information in policy making. Shakhnazarov concluded in 1974, "It is hard to establish goals in social policy without making a deep analysis of social groups and a tentative, preliminary assessment of their probable reaction to various measures."[28]

Elements of continuity and change are at play in the enhancement and adjustment of the CPSU's leading role under developed socialism. Historical continuity is reflected in the idea of a working-class party and its "vanguard" establishing political priorities. Soviet leaders' "consciousness" of the "laws" of historical progress guides the "spontaneous" development of society. But the working class is now part of an increasingly well-educated amalgam of proletarianized peasants, scientized laborers, and production-oriented intellectuals—all of whom live in a stratified but harmonious advanced technological society. Hence, the party must adjust its functions accordingly. Its self-imposed responsibility is to identify the common good, harmonize interests, shape national policy, supervise its implementation, and improve policy making and administration. That is, CPSU leaders are to consider the needs and opinions of organizations and strata, to adjudicate conflicts and claims for scarce resources, and to motivate bureaucrats and citizens to carry out programs that promote the welfare of all.

The governmental institutions—especially the councils of ministers, soviets, army, and secret police—are the other major sources of societal guidance in the USSR. Under Lenin the state bodies were vital instruments of one-party rule and of the proletariat's struggle with "antagonistic" classes that received military and economic support from abroad. Under Stalin the state bureaucracies, soviets, and CPSU were vital instruments of personal rule and of the proletariat's struggle with real and imagined "antagonistic" elements, domestic and foreign. Because of the amelioration of class conflict and the creation of a "nonantagonistic" social structure, "the dictatorship of the proletariat" eventually became "the state of the whole people." This formulation was triumphantly enunciated by Khrushchev in 1961, ignored in the early Brezhnev years, quietly revived in the late 1960s, and sketchily elaborated thereafter. E. M. Chekharin, director of the Higher Party School of the CPSU Central Committee, observes, "The state of the whole people is the political superstructure of the economy of developed socialist society."[29]

Soviet theorists emphasize that the government must expand its involvement in socioeconomic development.[30] Effectiveness and efficiency in policy implementation are the key criteria by which the CPSU leadership assesses the accomplishments of the state. Progress is equated with the administrative rationalization of state agencies and of national and local soviets. As Kas'ianenko observes, "The centralization of current and long-range planning of economic and social development strengthens the state's significance as the unified center of management."[31]

Soviet analysts contend that state power is an increasingly "conscious, creative, and active factor" stimulating economic productivity as well as expanding and redistributing economic output.[32] Government agencies are to become "scientized" by making greater use of modern organizational technology and improved methods of planning, management, and production.[33] Innovative techniques and approaches are to be drawn from socialist and Western economic and managerial science. Soviet commentators expect that computers will produce better information for decision making and that the scope and efficacy of state activities will thereby increase. Also, ministerial officials are exhorted to eliminate the many barriers to scientific-technological advances and to the application of new technologies and management methods in production. Furthermore, the Supreme Soviet and local soviets and (to a lesser extent) the trade unions, Communist Youth League, and other public organizations are to play a more active part in assessing the social, economic, ideological, cultural, and ecological consequences of national policies and in drafting laws in selected fields. More and more government officials are to participate at various stages and levels of policy making, and more and more citizens are to participate in policy implementation. Hence, contemporary Soviet theorists equate democratization with the expansion of state activities and declare that "the withering away of the state and the emergence of communist public self-administration" are goals that can be achieved only in the "distant future."[34]

Summarizing the present-day characteristics of the Soviet polity, Chekharin affirms:

> Soviet experience shows that society can enter the stage of developed socialism without essential formal changes in its political system. In the main, the political system society had at the previous stage continues to function.
>
> This, however, does not mean that it remains just as it was. Socialism is a dynamically developing society. . . . That is why the work done by the CPSU, the socialist state, and the public organizations to improve the Soviet political system, in order to meet the needs of the building of socialism and communism, has a profound social meaning and significance.[35]

In a word, Soviet institutions are to be "modernized," thereby accelerating the democratization of the polity.

Sources of Change. Soviet theorists maintain that an advanced productive base and socialist production relations are the preconditions for "progressive" change within the framework of existing political institutions and social relationships. Socialism is maturing by enhancing its capacity to manage its own evolution. Soviet analysts depict a society in which the central guidance institutions use the empirical findings of the natural and social sciences to help conceptualize goals, formulate policies, and improve policy making and administration. "Science, which is playing a growing role in resolving the social problems of developed socialism, is not only providing an important general orientation to the practice of management but is being used more and more widely in the *making* of concrete decisions on different levels of social management."[36]

According to Soviet commentators, the leaders of an advanced socialist society apply scientific approaches to uncover and conform to "the laws of social development." Decisions are to be based on probabilistic generalizations about actual patterns of group and individual behavior, as well as on the "objective" trends of historical development discovered by Marx and elaborated in theory and practice by Lenin and his successors. Also, scientific leadership is to be exercised "subjectively" by identifying and promoting behavior that enhances socialist democracy and eventually establishes communism. Progress toward these ends is considered irreversible, but it can be spurred or retarded by political leaders.

There are pale contemporary counterparts to the Leninist view that historical forces will develop progressively *only* with "conscious" party intervention to subdue powerful "spontaneous" forces. Brezhnev's ideologists emphasized that the CPSU's knowledge, power, and experience must help to aggregate social interests and resolve nonantagonistic contradictions. These accomplishments, in turn, enable party leaders to clarify and express society's central values in rapidly changing times. Soviet analysts stress that scientific societal guidance rests on both "objective" and "subjective" factors. Objective factors include the general course of history, the interests of specific classes, and the opportunities and problems thrust forward by the NTR. Subjective factors include the leadership's knowledge of objective conditions and its will and capability to transform society.

Soviet theorists emphasize the need for better management of scientific-technological and socioeconomic information.[37] The regulation of information is viewed as a means of augmenting power, especially the party-state's capacity to mobilize the population and coordinate complex subsystems. Power is gained by accumulating and utilizing numerous types of information about current policies, political-administrative procedures, and domestic and international conditions; by formulating and implementing comprehensive and long-term policies; and by periodically reviewing decisions and decision-making practices and adjusting them to changing circumstances. Hence, Soviet officials value flexible and integrated decision-making procedures, differentiated approaches to problem identification and resolution, and heightened sensitivity or responsiveness to diverse environments. Such capabilities are viewed as sources of power, not of vulnerability, under mature socialism.

Soviet observers argue that progress is the byproduct of evolutionary societal guidance from above rather than of creative disequilibrium from below. A developed socialist society, like all industrialized societies, monitors and synthesizes its internal tensions and dislocations. Purportedly unique in the mature socialist phase is the elimination of antagonistic contradictions. But nonantagonistic contradictions between the environment and society, between the worker and intellectual, and between the new technological potentials and old technologies persist. In other words, social and economic developments continuously generate competing interests. The need to eliminate or cope with these conflicts justifies the strengthening of the central party and state institutions.[38] A decentralized society such as Yugoslavia is deemed inappropriate, even harmful, under conditions of the NTR. If nonantagonistic contradictions are left unmanaged, they might produce *antagonistic* contradictions, and socioeconomic progress would be seriously impeded.

Most Soviet officials and analysts do not consider individual party organizations' failures and shortcomings to be even a nonantagonistic contradiction. The CPSU as a whole is beyond reproach, and collective leaderships since Brezhnev's have legitimized themselves by claiming to manage contradictions. Presumably, a serious incapacity to guide society scientifically (in the many senses noted above) would decrease the party's authority. Nonetheless, Soviet conceptualizations of developed socialism underscore the importance of centralized leadership. CPSU spokespersons affirm that the rationality of history in the advanced stage is rooted in a polity's capacity to combine scientific, technological, economic, *and* social progress.

The Soviet Concept of Progress

Soviet theorists stress that scientific and technological changes are necessary, though not sufficient, preconditions for social and economic progress. They are keenly aware of the reciprocal influences of scientific-technological, socioeconomic, and political-administrative developments. The NTR is to be welcomed, and to do so the institutions and activities of the maturing socialist society are to be adjusted. Let us examine some key Soviet ideas about progress and their theoretical and practical implications.

The Changing Role of Science. One finds a high degree of consensus in Soviet theorizing about the physical sciences' accomplishments during the second half of the twentieth century. Science has become "a direct productive force" in industrialized socialist and capitalist societies,[39] is beginning to "lead" production,[40] and is "transforming" nature.[41] Science is to be "mediated" through technology or the "materialization" of knowledge.[42] Before the NTR, the generation and application of technology were largely independent of science. Decisions about how to satisfy material needs through the creation of new tools and processes were based chiefly on technique or know-how. With the coming of the NTR, scientific knowledge molds technology. In socialist countries, "conscious" direction of scientific and technological development promotes the common good. As S. Tiushkevich notes, "The essence of science and its basic content are expressed in discovering the laws of development for the objective world and their use in the interests of society."[43]

Soviet theorists assert that science and production have passed through three increasingly interrelated stages.[44] The first stage extended from prehistoric times to the industrial revolution of the eighteenth and nineteenth centuries, during which science was alien to production and was largely an individual activity. The second stage extended from the appearance of machine production to the mid-twentieth century, during which scientific principles helped to analyze existing machines and to create new machines fulfilling immediate production needs. In the current third stage the relationship between science and technology is thought to be reversed, with science greatly influencing technological products and processes.

The growing role of science in production links theory and practice in dramatic new ways, Soviet commentators argue. A "science-technology-production cycle" is to be created because "science and technology are becoming so closely connected that they form a unified process, including man's cognition and utilization of the substances and forces of nature."[45] Science generates knowledge about the regularities of physical and social behavior, making possible the testing of reality and the production of tools through which people can control many aspects of their socioeconomic and ecological environments. In short, the creation of a science-technology-production cycle in some industries and its improvement in others are a central challenge to mature socialist societies.

Soviet analysts stress that modern science is the sine qua non of economic and social development, penetrating all managerial relationships from the political leaders' selection of modernization strategies to the organization of labor in the work place.[46] Party spokespersons and physical and social scientists underscore at least six important characteristics of post–World War II science.

First, pure research and its practical applications are meshed. "Science, as an activity for the production of knowledge, becomes the leading factor in the development of material production, defining the tempo of its growth and dictating the character of technological, organizational, and structural changes."[47]

Second, scientific advances are generated by interdisciplinary approaches and by syntheses of findings in diverse physical and social sciences. "Each new science in our time builds bridges over the abyss of a 'no man's land' by forming linkages between two or more sciences," asserts Volkov.[48]

Third, diverse sciences focus on human development. "The central problem of the natural, technical, and social sciences is the problem of man. In man are combined mechanical, physical, chemical, biological, and social processes and laws in their highest synthesis, and the modelling and determination of the physical and intellectual functions of the laborer are increasingly shaped by the tasks of the technical and applied sciences."[49]

Fourth, science shifts from an "extensive" to an "intensive" phase of development during the NTR. The numbers of scientists and scientific institutions become less important than creativity and productivity. Priorities and scarcities become more important too. The skillful management of scientific talent, equipment, and information is a critical political-administrative task.

Fifth, science is becoming a basic element of culture. Science is altering the structure and content of thinking, of perceived options, and of choice. New scientific ideas and new data have a growing influence on human values, attitudes, and beliefs and, in turn, on behavior.

Finally, science is altering the orientations and activities of diverse organizations and groups. According to Volkov, "Social institutions become dependent in varying degrees on the successes of scientific research activity," and "the methods and forms of activity in the various spheres of social production are transformed in conformity with the demands of science."[50]

Soviet theorists, then, concur that contemporary science is qualitatively changing all of the other productive forces as well as enhancing and integrating their capabilities. Fundamental research creates the concepts, hypotheses, and theories that serve as the foundation for new and improved products, materials, and technological processes. Applied research develops concrete ways to use the innovative ideas generated and confirmed by pure research. Experimental construction work and extensive testing produce the prototypes and new techniques that introduce scientific principles into production. And industrial and agricultural studies identify efficient ways to mass-produce the prototypes and to disseminate the techniques emerging from the earlier phases.

Soviet analysts recognize that a society's capacity to innovate is becoming a decisive factor in historical progress and that developed socialist societies must compete more successfully with highly industrialized capitalist nations. Progress is increasingly defined with reference to creativity and dynamism in the scientific-technological sphere and to the capability to incorporate discoveries and inventions into an economy, society, or polity with a minimum of short- and long-term "negative" effects. The harmful consequences cited are usually ecological but also include cultural, social, economic, and political repercussions.[51]

Furthermore, Soviet officials and theorists emphasize that managing the NTR's deleterious and unpredictable side effects is an ongoing, complicated, and collective responsibility. The resolution of "problem situations" is rarely complete or permanent, and decisions cannot be made once and for all. Khrushchev's exuberant and iconoclastic support for dramatic and ill-prepared new programs, his outbursts against bureaucrats, and his exhortations to the masses gave way to Brezhnev's regularizing of policy-making and implementation procedures, lowering of production targets and consumer expectations, and striving for durable cooperative relationships among professionals. Especially in the early 1970s, the Brezhnev administration persistently looked for more empirically grounded approaches and for technically competent personnel to help manage interrelated and recurring policy problems.

Brezhnev's collective leadership recognized that a decisive change was taking place in humankind's development, especially in the need to apply more scientific knowledge rather than more physical labor in the production of goods and services. This change was thought to necessitate some qualitatively different forms of management and considerably improved social and economic planning. New approaches were to combine scientific-technological innovations with feasible socioeconomic goals and with a rapidly evolving social structure. "The progress of science, technology, and production continuously places people, directly or indirectly, in situations that demand a complicated but at the same time effective managerial apparatus and *constant, flexible intervention in the affairs of society*."[52]

Soviet theorists view the planning and management of science as a decisive challenge to the leaders of all industrialized nations. Science has become vitally important to the organizational and cultural life of advanced society. But modern capitalism is thought to devote much of its scientific talent and resources to the development of military technologies. Soviet commentators contend that the West's sophisticated weaponry, together with the "imperialist" ideas and foreign policies it makes possible, compel socialist societies to expend scarce resources to develop comparable technologies. Socialist countries thereby ensure the safety of their people but fail to attain the quality of life they otherwise would have achieved if it were not for the allegedly retrogressive and exploitative nature of capitalism, which furthers the interests only of the bourgeois class. To be sure, Soviet observers acknowledge that scientific achievements have generated a high level of technological development in the West, and they stress that scientific-technological progress has produced a profitable and stable relationship between the largest corporations and political institutions at the national, state, and local levels. But "state-monopoly capitalism" blocks the types of planning and management essential to the public welfare.

Updating Marx's well-known theory of the immiseration of the proletariat under capitalism, present-day Soviet theorists suggest that the benefits of modern science will be distributed among increasingly few institutions in capitalist society. The implication is that the deprived or manipulated masses eventually will recognize the source of their powerlessness and will acquire (through largely unspecified means) the ability and determination gradually to overcome their highly organized, well-financed, and politically influential oppressors.

Soviet spokespersons have yet to produce a modern *Leninist* response to this distinctively Marxist argument. If such a response emerges, it will probably be based on an empirically grounded assessment of the power of important structures and strata in Western societies. Particularly important are the relative power of groupings and social movements vis-à-vis the formidable established corporate interests and the degree to which competing sides can influence government policy, economic performance, and public opinion. Also, contemporary Soviet theorizing might include a conservative variation of Lenin's famous views on "trade union consciousness"—that is, a call for heightened authoritarian leadership to offset the recalcitrance and inertia of a public incapable of understanding and pursuing its own best interests and preoccupied with "technological virtuosity"[53] and "commodity fetishism."[54]

Soviet centrists are less concerned about baneful Western influences and more realistic about the probable durability of capitalist governments and economies. "Capitalism is unable to cope with the NTR *completely*. Its endeavors to adapt [will] *ultimately* exacerbate the contradictions of capitalism even further. In order to meet the needs of the NTR, a transition from capitalism to socialism must be made."[55] This official interpretation is vague and its time frame is indeterminate. Nonetheless, it may be serving Soviet interests well in the present period of competitive cooperation *and* conflict among industrialized socialist, capitalist, and Third World countries.

Progress Disaggregated. The Soviet understanding of mature productive forces is rooted in its analysis of advanced capitalist society. Soviet ideas about material progress have been shaped by vocal condemnation and quiet admiration of the NTR's impact on Western countries. Modern capitalism is acknowledged to have accelerated economic growth and productivity. But capitalist production relations are thought to prevent the full development of the NTR and to preclude the scientific management of society. The NTR's unfolding in highly industrialized Western nations is viewed as an intensifying and irreconcilable contradiction. Material progress and social decay proceed simultaneously.[56]

Soviet commentators insist that the decisive difference between advanced capitalism and developed socialism is the "spontaneous" versus the "conscious" character of the institutions stimulating the NTR and distributing its accomplishments. The socialist state can master the NTR; the capitalist state cannot. Under developed socialism the NTR is channeled in directions chosen by leaders responsive to the public, shaped by national planning agencies and technical specialists, and carried out by state and public organizations. Hence, historical progress is correlated with the socioeconomic benefits derived from the fusing of the NTR with socialist forms of planning and management. Developed socialism is to increase the people's standard of living, the equitable distribution of wealth and public services, and the harmony and fairness of a social order grounded on mutual trust between the CPSU and citizenry.

Soviet observers perceive a disaggregation of progress in contemporary capitalism. On the one hand, an upsurge of scientific and technological power is inherently progressive. On the other hand, repressive capitalist production relations ensure that material progress will lead to social disintegration. Party spokespersons maintain that the basic contradiction of advanced capitalism is between the increasing scientific and technological potential to meet social needs and the decreasing governmental and corporate efforts to do so. They argue that the resolution of this contradiction is impossible under advanced capitalism and is the central challenge of developed socialism. In other words, progressive social aims and material capabilities are to be linked. Whereas the NTR develops in a dangerously anarchical and exploitative manner under capitalism, socialism is to harness the NTR's remarkable power to serve the common good.[57]

In essence, the Soviet conceptualization of developed socialism rejects advanced capitalism in its social domain but imitates advanced capitalism in most material domains. The Soviet image of progress is based on a favorable evaluation of advances common to both developed socialist and advanced capitalist societies. Although the NTR is a global process strongly influencing and influenced by contemporary capitalism, Western scientific-technological achievements are thought to be taking place in an antiquated socioeconomic and political-administrative context. The irrational purposes and values of capitalism are subverting the NTR's inherent rationality.

Soviet analysts stress that developed socialism provides the collective goals and institutions capable of stimulating the NTR's positive characteristics. Mature socialist nations contain perfectable organizational structures that ensure both the expansion of the productive forces and their fruitful uses. In advanced capitalist countries, progress is obstructed by the private ownership of the means of production, corporate pluralism, and a widespread unwillingness and inability to serve the national interest. Soviet commentators consider developed socialist societies to be the sole vehicles of progress in the contemporary era. Control over modern science and technology is viewed as the negation of capitalism's negative use of the NTR. Hence, *developed socialism is a negation of a negation*.

There is considerable ambiguity in Soviet perspectives on the nature of progress and on the means of fostering progress under developed socialism. This ambiguity is a by-product of Soviet views on the NTR's universality and on the NTR's effects in different socioeconomic systems. Soviet theorists affirm that most elements of the NTR are global in character, but they insist that the NTR requires decisive change in the essence and structures *only* of capitalist nations.[58] The material and human productive forces associated with the NTR are thought to be identical or quite similar throughout the world. But the impact of these new productive forces on the production relations and superstructures of modern societies is seen to vary enormously.

Soviet analysts draw the sharpest of distinctions between capitalist and socialist production relations. To be sure, the NTR does not affect all advanced societies in identical ways, and technological determinism can be criticized on many counts. But the capitalist/socialist dichotomy is undermined by the Soviet emphasis on the universality, power, and dynamism of the present-day scientific and technological productive forces. Emphatically denying any similarities between socialism and capitalism, Soviet spokespersons dismiss or deemphasize the similarity of *some* of the NTR's effects on all industrialized countries.

Soviet political leaders and theorists insist that the conceptualization of a modern industrial society is a task that has been resolved by *current* definitions of social identities. For example, A. S. Akhiezer rejects the use of "old, archaic methods of decision making" and notes the rapidly changing "norms and values" influencing advanced societies. But Akhiezer affirms, "The scientific and technological revolution finds a suitable social structure in socialist society. And socialist society finds in the scientific and technological revolution the scientific and technical basis for its progressive development."[59] Furthermore, Soviet officials claim that complete harmony exists between the USSR's historical legacies and the NTR. This claim is a conscious or unconscious denial of the necessity for qualitative change in those historical legacies. The core symbols and institutions of the Soviet polity are thought to be capable of meeting the challenges posed by the global NTR. The chief symbol of the political system's rationality—namely, the capacity to plan—and the organizations that give expression to this symbol are deemed to be socialism's "main advantages" in the era of the NTR.[60]

Soviet observers repeatedly assert that centralized planning is essential to scientific-technological and socioeconomic progress and that the Soviet state, unlike capitalist states, plans effectively and has the potential to plan even more effectively. These major assumptions impede critical dialogue within Soviet society about the nature of the political-administrative and cultural changes that may be required to spur scientific-technological innovation and socioeconomic development. According to Soviet analysts, the NTR necessitates planning that is comprehensive (i.e., capable of incorporating scientific, technical, economic, social, and environmental factors *and* their interconnections), prescient (i.e., capable of anticipating the course of the NTR and its long-range consequences), and scientific (i.e., capable of being grounded in a dynamic unity between Marxist-Leninist objectives and the social and natural sciences).[61] However, such planning is predominantly technical rather than strategic and is based on the premise that the political leadership can enhance its policy-making power through incremental organizational changes and modern telecommunications and computers.

Soviet presuppositions beg larger questions: If the NTR is producing radical change in the productive forces of all advanced nations in an increasingly interdependent world, how can the CPSU preserve the USSR's particular historical identity without at least a few fundamental changes? Why does the party leadership resist institutional reforms when resistance is probably diminishing both its authority and the USSR's scientific-technological and socioeconomic accomplishments? If Soviet commentators employed the capitalism/socialism dichotomy merely as an ideological justification for existing structures of power, or if they defined the USSR's place in history in isolation from universal history, then these theoretical ambiguities and their practical implications would be minimal. But Soviet leaders and theorists do not conceptualize progress in this way.

Instead, Soviet analysts underscore the scientific and technical breakthroughs taking place in capitalist societies. They affirm that technological innovations are peripheral to the essence of capitalism but are key components of its growing material and human productive forces. Moreover, they proclaim that all capitalist innovations are compatible with socialist production relations in principle and that many are useful in practice. Assessing the NTR's effects on education, V. Turchenko articulates this perspective:

> One must approach the problem of competition between socialism and capitalism in the area of education as one approaches competition in the area of technology: first, carefully study the development of education in bourgeois countries, use everything really progressive that furthers the labor productivity of teachers and students; second, resolutely expose and excise everything that is reactionary and antiscientific in the theory and practice of bourgeois education; third, on the basis of critical analysis and generalization of world practice, ensure the choice of the most promising and decisive tendencies in the development of public education, in order with the least expense to achieve the most results possible and to obtain superiority in this area in all basic aspects.[62]

Similarly, Soviet theorists exhort specialists to study the theory and practice of Western management and then to cull, adapt, and apply selected technologies and methods. Gvishiani maintains that a socialist science of management can make good use of Western cybernetics, systems analysis, information theory, and econometrics. Although Gvishiani affirms that American organization theory supports capitalist production relations and "performs a definite ideological apologetic function," he acknowledges that this theory investigates "a real process" and with an "effectiveness [that] has to a certain degree been proved by practice."[63]

Gvishiani emphasizes that foreign technology and expertise can help to develop the Soviet economy, but he warns that such experience must not be indiscriminately copied. It is necessary to take "a strict critical approach to foreign experience and the creative assimilation of its positive elements, and at the same time resolutely reject anything engendered by the distinctive features of capitalism and therefore inapplicable to socialist development."[64] Because socialist and capitalist production relations are qualitatively different and because management operates differently in the two systems, "exaggeration of the role of common features is both theoretically unsound and leads to practical mistakes."[65]

In short, Soviet analysts urge that advanced capitalist experience be judiciously borrowed and skillfully adapted so as to enhance the socioeconomic advantages of mature socialism. But party leaders have offered only the most general guidelines on how to identify the foreign achievements that will strengthen or weaken the basic features of the Soviet system. Gvishiani, following a rich Marxist tradition, distinguishes between the capitalist "form" and the socialist "content" of various technologies. Nonetheless, this distinction is subjective and its application to particular cases is quite arbitrary.

It is very difficult to discern the cultural components of foreign management approaches and technologies and to estimate their effectiveness, efficiency, and socioeconomic consequences in the USSR. Even if such analysis and prediction were simple, the decisions about which methods and equipment to import and how to adapt them would constitute highly *political* judgments. Moreover, CPSU spokespersons differ about the changes needed in the economy, society, and polity and the technological means of promoting these changes. Indeed, top party officials may agree only to experiment with certain innovations and periodically to evaluate how they are faring in practice. Assessments of opportunities and consequences are also politicized.

Most Soviet theorists do not confront the intellectual and practical ambiguities inherent in these realities.[66] Instead, they closely link the NTR's positive elements to the maturation of socialism and the NTR's negative elements to the irrationalities of capitalism. To be sure, some Soviet analysts acknowledge that the NTR is producing or could produce (in declining order of probability) harmful ecological, economic, social, and political consequences in socialist countries.[67] But the USSR has participated more and more in the international division of labor and has become more permeable to external influences. These developments have exposed Soviet society to problems and opportunities comparable to those of industrialized Western societies, enhancing the mutual benefits of scientific-technological, economic, and strategic arms limitation agreements between capitalist and socialist nations. Such commonality and parallelism suggest that if the universal NTR makes changes in historical legacies necessary in the West, then comparable changes are needed in the USSR. Many of the processes and products of interdependence contain or promote shared values. Whether these values support or subvert existing values is, of course, a moot question.

Soviet analysts deny the similarity of socioeconomic change in advanced societies, and they brand as "convergence theorists" all who suggest the universality of the potentials and pitfalls of scientific-technological development.[68] In turn, convergence is characterized as a weapon in the arsenal of imperialist propaganda and as a threat to the USSR's national identity.[69] Soviet commentators reject the idea that a mature socialist society cannot control the use of scientific and technological breakthroughs. Runaway science and technology, especially in the military sphere, are attributed to the private ownership of the means of production and to the absence of economic and social planning.[70]

Soviet ideologists assert that advanced capitalism creates all of the NTR's harmful elements. They claim that the Soviet system can winnow out virtually all of the NTR's negative features while assimilating its positive features. Likewise, CPSU leaders conceptualize progress with reference to scientific-technological developments in the West, but they refuse to acknowledge pressures for socioeconomic and political-administrative change similar to those occurring in the capitalist countries. They recognize that economic cooperation with the West must be mutually beneficial and based on complementary capabilities, but they repudiate the idea that Soviet and Russian legacies must be altered in response to universal dilemmas. Party spokespersons thus acknowledge the convergent changes in material and human productive forces but reject the possibility of convergent changes in historical identities.

Problems and Opportunities. Soviet theorists affirm that the NTR creates strong pressures for the transformation of capitalist and socialist societies and that both political orders face the challenge of combining the NTR with social progress. But they argue that developed socialism can produce a future production revolution, whereas advanced capitalism cannot achieve a production revolution without a prior social revolution. Only in a mature socialist society can the NTR fulfill its potential to serve the public interest. The crucial element in the Soviet rejection of the idea of convergence is the bifurcation of the historically "rational" from the "irrational." Socialism is a rational system; capitalism is an irrational system. The latter can become rational only by becoming isomorphic with the former.

The Soviet perspective is well summarized by N. Pilipenko:

> The superficial similarity of the phenomena of the NTR taking place under capitalism and socialism does not prove their "convergence." The current NTR is developing on the basis of highly industrialized production, a considerable growth of its intellectual potential, vast expansion of the front of scientific research, practical utilization not just of individual scientific discoveries, but of the cumulative result of the development of natural, technical, and social sciences, and their penetration into all spheres of production, management, control, culture, and sociopolitical and intellectual life of society. *At the same time, the content and especially the social consequences of the NTR are not identical in socialist and capitalist societies.* Whereas under capitalism it is used as a means of maximizing profits and bolstering the power of the capitalist class, under socialism its development helps to discover new sources of raw and primary materials, etc., essential for the growth of people's well-being and all-round development. Only under socialism does the NTR serve as a means of attaining the goals of social development that conform to the interest of the whole of society.[71]

Party leaders encourage different views on how to adapt the essential features of the polity, society, and economy to the NTR and on how to augment the authority of the state. But politial debate among the Soviet elites is directed toward the technical shortcomings of the present system rather than toward its basic characteristics or general course of development. According to a major Soviet text, "The differences of opinion in socialist society do not touch the fundamental issues of ideology; they are concerned with certain as yet unresolved questions of social life that require discussion."[72] Some Soviet officials and citizens, however, have found it difficult to accept the propositions that all "fundamental issues of ideology" have been resolved and that all the "unresolved questions of social life" are not fundamental. Hence, in the purported interests of collective and individual well-being, CPSU leaders since Lenin have strongly opposed unfettered dialogue between the party-government and the people. And controls over political discourse have enormous implications for the theory and practice of scientific-technological and socioeconomic development.

Marx emphasized that the productive forces of a society exert a much greater influence on the production relations and superstructure than vice versa. Following Lenin and Stalin, many contemporary Soviet theorists retort that the USSR's superstructure and production relations are having a greater influence over the productive forces than vice versa. Soviet analysts increasingly dispute these issues because of the NTR's dynamism and the cumulative impact of scientific knowledge and other "direct" productive forces. Whereas Soviet conservatives acknowledge the need for very gradual and controlled change in the USSR's superstructure and production relations and call for a "production revolution," Soviet reformers affirm that the NTR will eventually generate "radical" changes in production relations *and* the superstructure of Soviet society. Both interpretations are buttressed by little empirical evidence. Soviet conservatives and reformers maintain that the production relations and superstructures of capitalist countries are being "fatally" undermined by the NTR. But conservatives anticipate the swift demise of capitalism, and reformers anticipate its unhurried demise. For the conservative, the NTR is capitalism's poison; for the reformer, the NTR is capitalism's life-support system.

The Khrushchev and Brezhnev administrations expanded the USSR's involvement in the international economy and sought to control the accompanying multifaceted effects on Soviet society. Khrushchev destabilized and Brezhnev restabilized the Soviet political system, each for the purpose of promoting socioeconomic and scientific-technological progress. A Brezhnev successor could easily contend that major institutional and policy innovations are needed to meet unprecedented challenges and circumstances. He could maintain that the Soviet "superstructure" must be reformed so as to spur rather than retard a rapidly evolving "base." And he would probably stress that the economic base consists of myriad international and domestic components functioning in fluid and interconnected global and even outer-space environments.

The present and future Soviet leaders, then, are faced with an ideological and practical dilemma. On the one hand, they could *reject* the contention that the NTR thrusts forward problems common to all advanced societies. But, if a socialist society can "master" contemporary productive forces that it has played little or no part in generating, why cannot a capitalist society at the very least cope with, if not master, the same productive forces that it has largely or single-handedly generated? Also, the West is to date the chief producer and consumer of the NTR's benefits, which calls into question the structure and functioning of the Soviet economy and the official ideology that has long legitimized the CPSU and "self-legitimized" the power of its top leaders. Other sources of legitimacy are being developed such as consumer goods, social services, and "scientific" policy-making and managerial procedures. But these newer legitimation claims entail considerable short- or long-term risk because they are founded on the *performance* of the party and state rather than on symbolic, cultural, or patriotic appeals. The obstacles to improved economic performance are formidable, and they will probably remain so until major changes are made in the highly centralized economy or until a smoothly operating science-technology-production cycle is established.

On the other hand, if Soviet theorists *supported* the contention that the NTR thrusts forward problems common to all advanced societies, they could much more easily reiterate a seminal Marxist idea—namely, that a society's productive forces have a decisive influence on its production relations and superstructure and that the weaker reciprocal influences operate as well. This argument is bolstered by the international character of many scientific and technological advances and economic opportunities and problems (e.g., nuclear energy and inflation); the similar political-administrative and engineering options in some areas (e.g., trade-offs between environmental and industrial needs); and the inability of Soviet research and development to generate and apply as productively as the West key elements of the worldwide NTR (e.g., computers). But Soviet leaders focus on the *relative* technological levels of the USSR and of the advanced Western nations. By 1970 Brezhnev's collective leadership concluded that the ability to duplicate or adapt Western technologies was of even greater importance than the ability to produce indigenous technologies. Also, Brezhnev and other centrists decided that the political, economic, and military costs of selective cooperation with the West outweighed the benefits of self-sufficiency.

A small but increasing number of Soviet analysts argue that the NTR *is* producing common or parallel dilemmas in advanced socialist and capitalist systems. They quickly add that the Soviet polity and society possess vastly superior capacities to avoid or to deal with the negative elements and effects of the NTR. Recall the frank comments by the geographer I. Gerasimov on Soviet environmental problems and the terse warning by the political analyst Iu. E. Volkov about the bureaucratic and technocratic tendencies the NTR produces under both socialism and capitalism.[73] Observations such as Gerasimov's are becoming more and more common in Soviet writings, but assessments such as Volkov's are unusual. For both intellectual and practical reasons, Soviet theorists find it difficult to acknowledge global political-administrative problems, moderately difficult to acknowledge global socioeconomic problems, and less difficult to acknowledge global scientific-technological and ecological problems. Soviet analysts' keen appreciation of the politicization of these problems impedes analysis of the challenges confronting all highly industrialized nations.

Under Brezhnev, Andropov, and Chernenko, party leaders and social theorists emphasized that historical progress is to be achieved through the already established Soviet political structures and the expansion of their capacity to manage the NTR and the contradictions it produces. Not to be questioned are the purposes of the CPSU leadership, the existing institutions, and the basic political-administrative relationships. Soviet citizens are to criticize only the unfulfilled potential of individual CPSU, state, and public organizations to amass even greater authority and to use it to transform native and imported technologies into economic and social advances. The enhancement of a socialist polity's capabilities is deemed to be historically rational, because the domain of the irrational has been consigned exclusively to capitalism.

Soviet analysts conceptualize progress in terms of improved political guidance of all domestic and international aspects of societal development. The stated purpose of such guidance is to promote the material and spiritual well-being of the USSR's entire population. The utopian social objectives of Marx and Lenin have become less and less salient but were still operative aims of party leaders as diverse as Khrushchev and Suslov. At the outset of the Brezhnev period, *economic* goals seemed to have taken priority over *social* goals, and economic advances were expected to mitigate social problems. Only in the latter Brezhnev years did Soviet commentators publicly discuss the worsening of certain socioeconomic problems and the creation of new ones. Only then did they emphasize the linkages between economic and social development as well as the need to integrate economic and social planning and management.[74]

Throughout the post-Stalin era, Soviet theorists have averred that CPSU leadership is to be improved under evolving domestic and volatile international conditions in which authority and power can be easily devolved or dissipated. Also, Soviet analysts acknowledge shortcomings in the polity's shaping of human behavior and the natural environment, and they occasionally refer or allude to the uneven distribution of economic, social, and bureaucratic power in the USSR. But Soviet commentators argue that such dilemmas and limitations can be overcome as political-administrative rationality increases and as progressive scientific, technological, economic, and social trends develop. Soviet officials and specialists are thus keenly interested in decision making and implementation during the NTR, and we will now examine their ideas and actions in these important spheres.

3
"The Scientific Management of Society": Decision-Making and Implementation Challenges

Leading Soviet officials and specialists understand well the theory and practice of the "scientific management of society" (NUO).[1] Some Soviet analysts, like most Westerners, perceive a gulf between political-administrative ideas and experience in the USSR. The Soviet dilemma, as Paul Cocks affirms, is to bridge "the gap between policy design and implementation."[2] In order to eludicate this dilemma, we will compare the theory and practice of three interrelated elements of Soviet management—"democratic centralism," decision making, and administration—and present case studies on the management of complex systems and of research and development.

According to Soviet theorists, the sine qua non of developed socialism is "scientific management," whose crucial component is "scientific leadership." Top CPSU officials are to guide society as a whole and its component parts, comprehending and fulfilling the "demands" of the objective laws of social and economic progress. Such understanding and action make possible the maturation of socialist production relations and the utilization of maturing global productive forces. To manage society scientifically is to discover its progressive tendencies and to nurture them by means of planning, organization, regulation, and control. Afanas'ev calls for the timely disclosure and removal of the contradictions of socioeconomic development, ensuring "the assimilation or neutralization of internal and external disturbances and continuously perfecting the structural and functional unity of the system."[3]

Political leaders who respond to the "law-like regularities" underlying historical development are to accelerate social progress by bringing human subjectivity into conformity with the requirements of objective circumstances. Scientific management is possible only if party officials know both the objective laws and the subjective potentials of human activity and the "dialectical" relationships between them. "Absolutizing the objective conditions leads to fatalism, spontaneity, and the negation of the possibility of conscious influence on social processes. Absolutizing the subjective factor leads to voluntarism, subjectivism, highhandedness, and voluntaristic methods of management and is tantamount to a rejection of the scientific management of society, which presupposes cognition, calculation, and the use of objective laws and tendencies."[4]

The CPSU is to play the "leading role" in the management of a mature socialist society, "organically combining" modern science and technology with politics and administration. But Afanas'ev declares that "not even the best qualified central body is able to provide detailed management of a specific object,"[5] because the USSR's economic and social structure is becoming increasingly differentiated. A vast amount of reliable and timely expertise is to be infused into the policy-making process. The party must encourage specialists to contribute their know-how and suggestions in order to spur socioeconomic progress and to strengthen the political system. As those who are most competent and authorized to make decisions understand better the complex, interconnected, and dynamic conditions at home and abroad, arbitrary decision making gives way to scientific mangement.

In a mature socialist society, the top decision-making bodies are to take a "comprehensive" approach to societal guidance. By generating pertinent information from elites and ordinary citizens, CPSU leaders can expose and mitigate nonantagonistic contradictions. "The organs of management must be able to detect alterations in the system and act accordingly, so as to ensure not only its dynamic balance but also its continuous improvement and development."[6] Hence, social interests and needs are to be fulfilled; individuals are to think, feel, and act in conformity with the interests of society; and humankind is to be liberated from all forms of domination and exploitation.

So much for an introduction to the Soviet theory of scientific management. The practice of management in the USSR is quite different and has been well characterized by Western and Soviet analysts. Although the post-Stalin Politburos have been keenly aware of diverse political-administrative challenges, they have often lacked the will and capability to meet these challenges, especially after the mid-1970s. The problems and opportunities thrust forward by the NTR are mounting, however, and the Soviet polity's adaptive capacity is problematical. Four trenchant illustrations will suffice—the first two by Western commentators, the second two by Soviet commentators.

Cocks observes that "the significance of the USSR's belated awakening to the modern systems age is *the discovery by its rulers that the Soviet system is, in fact, not a system* in the sense of a well-integrated assembly of parts pulling together in the same direction. On the contrary, compartmentalization of activity and fragmentation of government stand out as dominant features of contemporary Soviet society."[7]

V. Zaslavsky maintains that CPSU leaders have an ambivalent attitude toward socioeconomic data and are especially reluctant to collect, let alone disseminate, data that may call into question current national policies, policy-making procedures, and power relationships.

> The attitude of the Soviet leadership toward concrete sociological research has always been a mixture of enthusiasm and distrust. This type of research provides relevant information about society, and it would seem necessary for the Soviet hierarchy to have this knowledge. A similar point of view was once expressed by E. Shils: "The truth is always useful for those in power, regardless of whether or not they wish to share this truth with the governed." Many Soviet sociologists have had similar thoughts. However, experience has shown that concrete sociological research can be harmful in at least two ways. The acquisition of objective information about society requires the creation of fact-finding machinery that is independent of ideology and has the freedom to collect information. Such machinery, not contained within the framework of a single ideology, becomes a constant political threat. Furthermore, the knowledge of their own failures can weaken the unity of the ruling class, increase competition within the group, and, in general, have a demoralizing effect.[8]

B. Z. Mil'ner contrasts the potential and actual achievements of Soviet economic management, and the differences are striking.

> Public ownership of the means of production eliminates fragmentation of the economy and turns it into a unified organism. There is now a real possibility of ensuring a balance among all sectors and all units of the national economy as an integral system.
>
> However, despite thoroughly specialized branch management, there is *frequently no coordination or dovetailing* of the activity of the various units, agencies, and officials who handle common tasks. This can be observed even within the framework of a single enterprise or association (not to mention a whole branch or group of branches). *Responsibility* for the introduction of new equipment, for the quality of output, for the completeness of deliveries, for the organization of

> evenly paced work, etc., is in many cases *undefined and scattered*. The problem is exacerbated a hundredfold when programs of all-union importance or programs for the comprehensive development of specific regions are involved, programs that must deal not only with production questions but also with questions of transportation, ecology, demography, cultural and consumer services, etc.[9]

V. A. Trapeznikov, a member of the Academy of Sciences and head of the Moscow Institute for Management Problems, compares the theory and practice of Soviet industrial administration and finds the latter sorely lacking.

> Technical progress is not simply new methods of production, machines, and technological processes. It also involves improving management at all levels. . . .
>
> A requirement of management theory is effective *feedback*—that is, the influence of succeeding segments of the management system on preceding ones. In the national economy, this includes the consumer's influence on the producer (which at present is negligible), the influence of results on management, etc. *In our economy, feedback is extremely weak and operates only with great delays*; when there is a scarcity of a particular product, feedback virtually disappears and output quality declines, since the rule "If you won't buy shoddy goods, someone else will" becomes operative. . . .
>
> Various *incentives* are known that channel the desires of individual people and collectives along paths that are expedient for the economy. One of them—which, unfortunately, is seldom used—is *competition and, on this basis, the selection of optimal decisions*. Constant comparison of the quality of manufactured goods with the achievements of other Soviet and foreign firms is an effective engine of progress. This is especially apparent in the defense industry, which makes continuous and inevitable comparisons with foreign equipment, comparisons that force us to maintain a high scientific and technical level for the articles we make.
>
> Instead of active competition, frequently a directly opposite path is taken, one in which a given organization acquires a *monopoly* in its field. The saying "We'll mass our forces at one point" is offered as justification. However, *a monopoly hinders technical progress and ultimately extinguishes scientific and technical initiative*. Then the customer has nothing to choose from—he's forced to take what's offered, and the organization that has the monopoly, gradually abandoning an active

search for new solutions, lives by the rule "If you don't like my product, make your own." A "lazy brains" syndrome is fostered, and technical progress is retarded.[10]

Changing the conditions described in these four illustrations is the chief goal of the would-be "scientific" managers of Soviet society. Their concerns range from macrolevel analysis of societal guidance to micro-level analysis of administrative behavior in local party, state, mass, and production organizations. Let us first examine the general and then the specific elements of Soviet management theory and practice.

"Democratic Centralism"

Lenin formulated the concept of "democratic centralism"[11] in 1906 when the Bolsheviks were still a faction of a radical socialist party struggling to overthrow the tsars. His portentous ideas were that the representatives of all party bodies should be elected; party bodies should periodically report to one another; party discipline should be strictly enforced by subordinating the minority to the majority; and the decisions of higher party bodies should be binding on lower bodies. Local and national initiatives were to be mutually reinforcing rather than mutually exclusive. Democracy and centralism were to be in "dialectical unity."

Ever since the 1917 Bolshevik revolution and even during the Stalin era, Soviet leaders and analysts have reiterated that democratic centralism is the cardinal organizational principle of party and state management and that democracy and centralism must be simultaneously enhanced. The post-Khrushchev collective leaderships have tried to delegate greater responsibility to the major bureaucracies and at the same time have guarded against the possible dissipation of the national party organs' power to initiate policies. Moreover, Brezhnev, Andropov, and Chernenko facilitated the upward and downward flow of information within the bureaucracies and society, while trying to preserve the Politburo's and Secretariat's capability to choose among policy alternatives supported by major institutional coalitions and social strata. Today's party leaders recognize that matters of secondary or merely local importance continue to clog the central policy-making mechanisms and that such matters could be decided at lower levels more effectively and efficiently on the basis of more timely and pertinent socioeconomic and scientific-technological information. But leading Soviet officials must weigh this potential benefit against the distinct possibility that national power will be further dispersed into weakly connected and self-serving departmental and local units. Especially since the late 1970s, bureaucratic and parochial interests have reduced the oligarchs' capability to implement domestic programs. General secretaries after Stalin have been frustrated by this trend but, with the fleeting exception of Andropov, have been unable to reverse it.

More and more Soviet spokespersons are addressing the key issue—namely, the need to increase elite and mass participation in the assessment and adjustment of decisions *after* they have been made, while curbing fissiparous tendencies in the growing party and state bureaucracies. Soviet officials must be circumspect because the party rules still proscribe discussion of political-administrative questions once a decision has been taken.[12] The authority to alter decisions gravitated long ago to the highest levels of the party. But the theory of democratic centralism ignores the ongoing nature of decision making and the need for continuous feedback to help reevaluate goals and, if necessary, to revise them.

Today's Soviet theorists stress that the nature of centralism and democracy, not merely the "balance" between them, are evolving. Afanas'ev declares: "In order to *strengthen* centralized and planned management, it is necessary to *redistribute* certain functions from top to bottom and from the center to the localities, over an increasingly wide circle of organizations and people."[13] Soviet commentators contend that deconcentrating decision-making power and encouraging initiatives from below (especially from the industrial and production associations) are essential to scientific management. That is, they recognize that the problem-solving capabilities of the central party organs can be *increased* by the delegation of selected responsibilities to administrative and territorial subunits.

Soviet analysts have linked the concept of democratic centralism with the democratization of a maturing socialist society. Shakhnazarov maintains that democratic centralism is the root organizational principle of the party, state, and "entire socialist political system" and is a "decision-making mechanism" only in a "narrow sense."[14] R. I. Kosolapov affirms, "The most effective correlation of democracy and centralism in the life of a socialist society takes shape gradually and is continuously corrected by practice, with *democracy always gaining additional ground*."[15] Hence, Shakhnazarov argues: "Democratic centralism is the optimum combination of the *interests* of the whole and its parts, of society and the individual, of the state and the citizen, of the center and the periphery, and so on."[16] Kosolapov concludes: "Determining and maintaining the right correlation of the democratism and centralism suiting the developmental level attained by society is one of the main aspects of improving the entire social system of socialism. In effect, this is the chief issue of political guidance in the new society."[17]

The jurist M. I. Piskotin analyzes democratic centralism in terms of political ends and means.

> The search for effective goals is more important than the search for effective means of achieving them. Indeed, this is the idea that lies at the basis of an "eternal" question of management—finding the correct relationship between centralization and decentralization in making management decisions. A centralized system of decision making is needed to ensure the selection of effective goals in any management structure. But, because this selection has an a priori and autonomous tendency, it leads to the separation of the formulation of goals from the selection of effective methods of implementing them and consequently makes necessary the distribution of the process of goal setting among elements of the management structure—that is, decentralization. Although it may seem paradoxical, all attempts to establish this division proceed from tacit assumptions about its stability and predictability. Hence, suggestions are needed about improving the mechanism for communicating information from the object to the subject of management, from the managed to the managers and back again.[18]

In Piskotin's view, a "compromise" between centralization and decentralization is reached "by *integrating* the analysis and evaluation of information and output derived from functional specialization."[19]

Piskotin identifies two groups of problems concerning democratic centralism. "The first consists of questions connected with guaranteeing the democratic nature of centralized leadership. . . . The second group consists of problems of combining centralization and decentralization. . . . In an intermediate position between these two groups lie questions concerning concentration or deconcentration, the distribution of decision-making responsibilities, and the degree of centralized leadership at various levels, including all-union, union-republic, and republic organs of state management."[20] According to Piskotin, the first set of problems includes the democratic structure of national institutions, the accountability of these institutions to the people, and the development of democratic forms and methods in political-administrative activities. The second set includes the rights and responsibilities of national and regional bodies; the combination of centralized leadership with the economic independence and initiative of enterprises, associations, and other organizations; and the distinctive characteristics of centralized and decentralized management in the sociocultural sphere.

The first group of problems is not well developed in Soviet theory, but Shakhnazarov, Kosolapov, Piskotin and others address them. The main benefit of democratic centralism, in Shakhnazarov's view, is that it enhances the CPSU leadership's capability to harmonize diverse social needs and interests; to produce unified national policies through consensual rather than confrontational policy-making procedures; and to combine "society's need for centralized guidance with the need for maximum initiative and selfadministration" of the myriad party, government, and production units. Shakhnazarov poses an important question: Can democratic principles be violated in a developed socialist society? His answer is no, and it rests heavily on the following assumptions:

> Centralism is the antipode not of democracy but of unlimited centralization. . . . There may be no strong, effective, and highly authoritarian central power, and local authorities may be highly independent in making their decisions. But this does not mean that democracy prevails. And, conversely, *a strong and highly authoritarian central power may be wholly democratic*. The democratic nature of the state depends, therefore, not on whether it is centralized or not centralized but chiefly on its planned character, on whether the central and local bodies of power are in the hands of the people and exercise the people's will and interests, and on whether the mass of the people participate in the work of these bodies and keep them under their control.[21]

Shakhnazarov argues that social, administrative, and legal controls ensure the democratic functioning of central institutions under mature socialism. Among the USSR's "antidotes" to bureaucratism and elitism are the abolition of privately owned land and corporations; the restrictions on transferring personal wealth to one's heirs; the illegality of using money to acquire political power; and the constitutional provision (Article 49) enabling citizens to submit proposals to state and public organizations and to "criticize shortcomings," as well as obligating such organizations promptly to reply to grass-roots proposals and to correct any shortcomings uncovered without "persecution for criticism." Also, Soviet citizens have the right to judicial review of "complaints against the actions of officials" and to "compensation for damages caused by illegal actions of state and public organizations" (Article 58). Are party organizations subject to such constraints, in theory if not in practice? Significantly, but ambiguously, the USSR Constitution (Articles 6 and 100) and analysts such as Shakhnazarov note that the party is both a public organization and the "nucleus" of all other public and state organizations.[22]

Shakhnazarov views the democratization of a socialist society as a difficult "*search* for optimum principles and institutions." This search is based "largely on *experimentation* and the objective progress of social practice" under complex and rapidly changing circumstances at home and abroad. CPSU leaders must strive to attain an "optimum balance of *competent administration by specialists and constant control over what they do by the mass of the people* who, with a high degree of political knowledge and tradition, *take an active part in making the basic decisions*."[23]

R. I. Kosolapov, having noted the importance of centralized leadership, observes, "The common regularities of socialist and communist construction operate differently in different places, and their optimum effect on the activity of nations, collectives, and individuals may differ *very considerably* from their effect on the scale of whole societies and much more on the scale of a world system." Because a differentiated approach is needed to conceptualize and respond to the problems and opportunities of a developed socialist society, "*cognized necessity can be put into effect only through democracy*, that is, the granting of broad independence to collectives and to individuals."[24]

Soviet theorists argue or imply that leadership rests on authority, not on coercion, and that greater elite participation in the politics of principles and greater public participation in the politics of details are important sources of legitimacy and effectiveness. Too much centralism will stifle the creativity of specialists and lead to arbitrary rule. Too much democracy will spur individualism, even anarchy, and seriously weaken the party-state's governing capabilities. Final decision-making power must remain concentrated in top political bodies in order to preserve the fundamental characteristics of the Soviet polity. But bureaucracies, collectives, and individuals must take more and more initiatives "*in accordance with the goals of communist construction . . . and for the purpose of strengthening and developing the socialist system*" (Articles 47 and 50 of the USSR Constitution, emphasis added).[25] Hence, only a shifting mix of democracy and centralism enhances mutual trust between the leaders and the led (the "subjects" and "objects of management") and furthers their cooperative pursuit of the public interest.

The second group of problems Piskotin identifies is receiving increasing attention from Soviet commentators and is the source of considerable disagreement. Afanas'ev asserts that the "optimal combination" of democracy and centralism "depends in the final analysis on the level of production, on the state of social relations, and on the specific historical circumstances." Afanas'ev points out that democratic centralism has distinctive characteristics in various "spheres of management" such as economic, social, and cultural life and in the party, military, and public organizations. Moreover, he affirms that the appropriate mix of democracy and centralism depends on the specific stage and level of the decision-making process. In Afanas'ev's view, the drafting of a decision and its verification [*kontrol'*] necessitate broad participation and discussion, whereas the final making of a decision and accounting [*uchet*] require a high degree of centralization and personal responsibility. Furthermore, he maintains that democratic centralism must be tailored to current domestic and international political conditions, to the nature of the problems to be solved, to the information processing and other capabilities of the institutions assigned to solve them, and to the time available for the task.[26] Hence, Afanas'ev takes a highly differentiated approach to democratic centralism and affirms that management organs must continuously adapt their decisions *and* decision-making procedures to changing contextual conditions.

The Brezhnev leadership encouraged such differentiated perspectives on the theory and practice of democratic centralism. According to Soviet analysts, the NTR necessitates greater *centralization* in the following policy and policy-making spheres: strategic economic and social planning, resource allocation, research and development, capital construction, wages, prices, credits, calculation of economic growth rates, and selection of indicators and incentives to stimulate economic productivity. The same Soviet analysts maintain that the NTR necessitates greater *decentralization* in other policy and policy-making spheres: economic management at the association and enterprise or farm levels, use of wage funds and profits to improve the quantity and quality of production, introduction of technological innovations in industry and agriculture, development of new revenue sources, clarification of the prerogatives and duties of subordinate units and personnel, economic cost accounting, and choice of methods to implement central directives.[27]

Soviet commentators disagree, however, about the specific spheres to be centralized or decentralized, and some of these differences are significant. For example, Afanas'ev contends that "decision making . . . demands centralization"; Shakhnazarov calls for mass participation in making "basic decisions" and identifies decision making as one of the most promising areas for "worker participation in management."[28] Piskotin minimizes the importance of precisely delineating administrative rights and responsibilities; Afanas'ev emphasizes its importance.[29] Mil'ner calls for the centralization of research and development *and* of "the application of scientific and technical achievements"; Piskotin observes that "one can hardly expect centralized measures to introduce scientific and technological achievements into practice unless production collectives are really interested in utilizing them."[30]

Diverse views about centralization and decentralization are a by-product of the Brezhnev, Andropov, and Chernenko Politburos' desire for scientific-technological and socioeconomic progress on the one hand and political stability on the other. One could argue that institutional change *or* continuity is especially important in the era of the NTR, and Soviet analysts have advanced both of these competing arguments. Because party leaders have not endorsed a uniform theory of the NTR and because many economic and social problems have worsened since the mid-1970s, Soviet disputes about the policy and policy-making implications of scientific-technological and socioeconomic developments are not surprising.

What *is* surprising are the occasional Soviet assertions about the harmful consequences if political-administrative changes are obstructed or delayed. Words and phrases such as *inertia*, *policy mistakes*, *philistine*, *toady*, and *abuse of power* appeared in the party press during the second half of the Brezhnev administration, and these criticisms were leveled (indirectly but unmistakably) at contemporary national policies, policy making, and individual policy makers (including, on rare occasions, Brezhnev himself).[31] Also, ominous hypothetical situations were discussed or noted by analysts such as A. P. Butenko, head of the Department of Political and Ideological Problems of the USSR Academy of Sciences' Institute of Economics of the World Socialist System. *If* a ruling party neglects the interests of the working people, it will lose their "trust" and "enormously weaken" its political power. *If* a party makes "gross mistakes" in its policies, "extremely dangerous consequences" will ensue, "no matter what historical stage of socialist construction is involved." *If* a socialist party and state fail to give adequate consideration to the interests, moods, and opinions of the masses, and *if* these factors are "ignored" or minimized, "the danger that the society's political organization might break away from the broad laboring masses arises." Butenko identifies as "crisis situations" the years 1968–69 in Czechoslovakia and 1970 and 1980–81 in Poland. But Butenko makes it clear that, without the appropriate kind of leadership for a particular phase of development, "political crises" can occur in any socialist society and in any developmental phase.[32]

In other words, Soviet theorists affirm that the effectiveness and legitimacy of party leadership depend on the feasibility of its policies, on the consultative nature of its policy-making procedures, and especially on the extent to which both respond to the objective requirements of the NTR and to the changing needs of the masses.[33] These needs, it is presumed, are best served by a one-party system. Butenko would criticize the middle-class intellectuals of the Prague Spring and the working-class Solidarity trade union movement as well as the reformist Czech and Polish party leaders, all of whose activities are thought to have jeopardized the predominant role of the indigenous parties and Soviet control over them. Lacking in both Czechoslovakia and Poland were "scientifically substantiated" policies combining traditional Soviet values and structures with the NTR, new management methods enhancing but regulating elite and mass participation in public administration, and ideological ties binding all organizations and citizens and restraining radical elements.

Leading Soviet officials are not anticipating political unrest of the kinds that have periodically jarred Eastern Europe since Stalin's death. But Soviet analysts have acknowledged that the CPSU can do a poor job of leading, even in the present-day "state of the whole people." Without scientifically grounded policies and close ties between the party and the population, "socialist construction may, at *any* stage, arrive at a *dead end* and pose the *greatest danger* to the fate of socialism." Butenko elaborates:

> Socialist society is characterized not just by certain more or less readily resolvable, nonantagonistic contradictions; it is still not free of acute conflicts and may encounter serious clashes. . . .
>
> The primary cause of negative phenomena in socialist society lies in the unskillful resolution of contradictions inherent in socialism. . . .
>
> In the course of the development of socialism, a situation may very well arise in which the improvement of production relations that is required in order to accelerate social progress is impeded by obsolete elements in the political organization of society and ossified management forms. . . .
>
> Consider, for example, the problem of the managers and the managed under socialism. Their interests are in contradiction, albeit *nonantagonistic* contradiction, with one another. However, if the managers become divorced from the managed and start to take advantage of their position to further their own selfish, group interests at the expense of the interests of society and the working people, these selfish group interests and the interests of the working people may become mutually exclusive and may take on, in this respect, the nature of an *antagonistic* contradiction. It should be especially stressed that this is by no means a case of "residual antagonism" or "vestiges of capitalism."

> This sort of backward, regressive evolution not only can occur but *has* occurred under socialism, as documents of the socialist countries' communist and workers' parties have noted.[34]

Hence, Soviet reformers argue that nonantagonistic contradictions may become antagonistic in a mature socialist society. Harmonious classes and strata may undergo "polarization," and "a social struggle between socialist and antisocialist forces" may ensue, creating the highly unlikely but real possibility of "the restoration of capitalism." Leadership "failures," not "the nature of socialism," generate such "political crises."[35]

Numerous political warnings and exhortations, as well as thoughtful sociological analysis, are contained in a remarkable statement by Butenko in 1981.

> The form of political power that evolves in a particular country during a revolution and then survives over many decades, absorbing within itself the unique national and governmental conditions, becomes traditional and difficult to change. This is a mixed blessing because administrative personnel are able to accumulate considerable experience and administrative structures are able to achieve considerable success, but implementation of the power of the whole people may be hindered to a certain extent. The conservatism of the previous experience, which is useful in its own conditions and which does not correspond to the new situation, hinders further development, as does the survival of obsolete forms of organization.
>
> Inasmuch as society's political organization develops with relative independence, preservation of its obsolete forms does not produce negative consequences immediately. At first it seems to preserve the means of implementing democratic centralism, the methods of management, and the forms of mass participation in public life that had seemed entirely effective just yesterday. But life marches on. Experiencing industrialization and collectivization, the undeveloped economy of the transitional period is superseded by a powerful socialist economy resting on the latest achievements of science and technology. Therefore, the forms of centralized leadership and the system of administrative levers that were previously acceptable become all the more unacceptable in a complex multisector economy. The earlier means of exercising democratic centralism, which evolved with a relatively small number of factories and plants participating in a system in which a single center plans the activities of each enterprise and provides it with the capital investments and other resources it needs,

cease corresponding with the new situation in which there are thousands of economic units, specialization and cooperation are multiplied many times over, and economic relationships have become more complex. The growing initiative of the masses and the utilization of local possibilities and resources are set against management methods that had recently been effective but have lost their virtues in the new conditions. *If the necessary changes are not made in time, yesterday's virtues of society's political organization become its shortcomings.* This requires constant improvement of the entire system of management forms and methods. Moreover, . . . *if obsolete forms of leadership and management of public affairs are not eliminated in time, the situation would become all the more intolerable in view of not only socioeconomic but also political development.* The forms of mass participation in the administration of public affairs would have been developed for the previous conditions; they would not reflect the new stage of development of society's political organization in socialist conditions, where *power implemented through the political organization must be power not only for all laborers but also through the laborers themselves.*[36]

Butenko is emphatically calling for innovative leadership and institutional adaptation. He is even more apprehensive than Shakhnazarov about nonantagonistic contradictions and obviously disagrees with Shakhnazarov's view that democratic principles cannot be violated in a developed socialist society. Also, Butenko would probably criticize the deterministic implications of R. I. Kosolapov's view that democracy "always" gains ground over centralism and would stress instead the participatory dimensions of scientific management under mature socialism. Butenko would very likely agree with Mil'ner that "centralism does not imply a rigid or detailed regulation of the activity of lower-echelon units and agencies, but coordination of the work of all sectors of the economy for the most rational and efficient solution of problems."[37] Significantly, Butenko stresses that the citizenry must increasingly participate in making "*final decisions on vitally important issues*" and that it must "directly" implement more and more government activities.[38]

Briefly stated, Soviet analysts in the post-Stalin period have emphasized the complexity, interconnections, contradictions, uncertainty, and dynamism of domestic and international politics, and some have cautiously suggested that power relationships and managerial practices should be modified accordingly. Brezhnev declared in 1981, "Democratic centralism is an immutable norm of the Communist Party's life. It presupposes, among other things, the closest possible link between the center and the localities, between the party's leadership bodies and the party masses. And this is a two-way link."[39] The general secretary's statement can be interpreted in diverse ways, and it was probably intended to legitimize differentiated approaches to democratic centralism. Throughout Brezhnev's collective leadership, however, Soviet spokespersons called for or discreetly advocated political-administrative changes that would adapt the theory and practice of democratic centralism to contemporary conditions. At the same time, the top CPSU leaders' claim to rule on the basis of a special knowledge of historical forces was gradually deemphasized or discredited.

Decision Making

Since the time of Lenin, the CPSU rules have contributed greatly to the centralization of decision-making power. Especially important are the rules obligating Communists to discuss policy alternatives freely in party meetings and in the press, but only *until* "the organization concerned has adopted a decision."[40] Uncertain about their authority to make decisions of various kinds, middle- and lower-level CPSU officials have looked to the center for specific directives, especially regarding the adjustment or abandonment of current policies. As Alfred Meyer inquires, "If the principles of democratic centralism forbid the criticism of agreed-upon policies, who has the right to suggest that any such policy has, perhaps, outlived its usefulness or should be reexamined in the light of changed circumstances?"[41]

Power is concentrated in the Politburo, Secretariat, and Central Committee departments and is exercised with extraordinary secrecy. Brezhnev, describing the almost weekly Politburo and Secretariat meetings between 1976 and 1981, offers a highly authoritative but cryptic account of national policy making:

> The questions submitted for the Politburo's consideration were carefully prepared beforehand. The range of these quesitons was extremely broad and multifaceted. Many of them are becoming more complex all the time. In some cases, the Politburo sets up special commissions for the comprehensive study of phenomena and the drawing of general conclusions, as well as for the taking of appropriate practical steps as quickly as possible.
>
> During the preparations for the meetings and during the discussions, various opinions were stated, which is quite natural, and numer-

> ous comments and proposals were made. However, all decisions were adopted in a spirit of complete unanimity. It is this unity that is the strength of collective leadership.[42]

Brezhnev observes that "*various opinions*" were expressed about "*key issues related to the practical fulfillment of the decisions* of the 25th CPSU Congress and the Central Committee's plenary sessions and *to new phenomena in domestic and foreign policy*."[43] Tacitly acknowledging vigorous debate on some issues, he asserts that "all decisions were adopted in a spirit of complete unanimity," not that "all decisions were adopted unanimously." But Brezhnev sheds little light on the politics of the reaffirmation, reassessment, readjustment, or rejection of national policies and policy-making procedures. In a word, the Soviet policy process is well shrouded.

How the Soviet political system formulates goals and priorities and how specialized elites express their different views and interests before *and* after the making of decisions are subjects that Soviet management theorists sidestep in their published investigations. Soviet commentators eschew detailed analysis of the institutional and policy-making changes that will accompany the NTR and its socioeconomic consequences in the USSR. Notwithstanding reformist views such as Butenko's, most party spokespersons assume that new methods will serve traditional purposes or that traditional methods will serve incrementally adjusted purposes. Democratic centralism is analyzed chiefly on a normative rather than descriptive basis, and it is always portrayed as the flexible but wise middle course between the Scylla of "anarchic decentralization" and the Charybdis of "bureaucratic centralism."

Soviet interpretations of decision making place considerable emphasis on properly defining the nature of a situation rather than on rewards and sanctions. Not linking motivations and interests with perceptions and expectations, and not underscoring that officials in different positions and roles are likely to view opportunities and problems in different ways, Soviet analysts affirm that deviations from national objectives are somehow pathological. For example, top CPSU officials have long complained about "departmentalism" [*vedomstvennost'*] and "localism" [*mestnichestvo*]. The Soviet leadership has always claimed to produce a clear and homogeneous interpretation of national aims, societal needs, and domestic and international conditions and has contended that competent and conscientious bureaucrats share these perspectives and act accordingly. Such presuppositions create a wide gap between normative theory and bureaucratic behavior. Equally wide is the gap between the party leaders' considerable experience in managing organizational conflict and the incorporation of this experience into Soviet decision-making theory.

Although Western organization theorists have traditionally stressed the inseparability of politics and administration, Soviet theorists have traditionally stressed the dichotomy between politics and administration. Since the mid-1960s, however, Soviet commentators have disputed the relationship between leadership [*rukovodstvo*] and management [*upravlenie*]. Afanas'ev flatly stated in 1968 that scientific leadership and scientific management were "identical" because they both comprised "the conscious influence of people on social processes." V. V. Semenov promptly criticized Afanas'ev in the most authoritative CPSU journal, *Kommunist*, tersely noting the importance of the party's ideological activities and implying that the party was primarily responsible for leadership and the state primarily for management. Afanas'ev immediately recanted and distinguished between the policy-making functions of the party and the administrative functions of the state, declaring that "*party leadership is the highest, most general, and essential element of scientific management.*"[44] But Afanas'ev, his political star having risen, asserted in 1979 that scientific leadership and scientific management

> lie on the same plane and are, generally speaking, *identical*, insofar as both the one and the other represent conscious influence on society or its separate elements on the basis of objective laws. Their overall purpose is also *identical*: ensuring the effective *functioning and development* of socialist society. *Both* management and leadership are aimed at revealing and making use of the advantages and possibilities of socialism and guaranteeing the successful building of communism. However, they *differ* as regards their agents, content, and, most important, the ways and means of bringing influence to bear on their subject.[45]

In contrast, F. M. Burlatskii, philosophy department head of the CPSU Central Committee's Institute of Social Sciences and vice-president of the Soviet Association of Political Sciences, argues that "the basic difference between leadership and administration is a difference in the extent to which power is delegated," and that leadership possesses more authority than do management, organization, and control.[46] Also, N. V. Chernogolovkin observes, "It is the task of management chiefly to ensure the optimal functioning of the object, whereas leadership is to ensure its development. The aim of guidance is thus to secure the progressive development of society toward communism, that is, to direct social progress along the path of transition from the lower stage of communism to the higher."[47] Chernogolovkin, much more than Afanas'ev, identifies leadership with goal setting, effectiveness, and ideological mobilization and identifies management with goal seeking, efficiency, and organizational mobilization. In a word, the relationship between scientific leadership and scientific management has not been resolved and was the subject of "lively discussion" (Afanas'ev's phrase) throughout the Brezhnev years.[48]

The Soviet debates about leadership and management have legitimized proposals for improving political-administrative processes and performance while at the same time preserving the power of central organs to utilize, adapt, or ignore such proposals. Especially important is the increasing attention devoted to decision making in the burgeoning Soviet literature on managing the economy and society. The various stages and levels of decision making, the role of experts and expertise, the sociopsychological components of leadership, and the utility of Western systems theory are examined in macro- and micro-level analyses. Afanas'ev affirms:

> The more complicated and hierarchical a system of management, the more difficult it is for personnel to communicate and assimilate information and the greater the possibilities for distortion and a poor understanding of decisions. This is one of the reasons why the number of levels of management in administrative systems must be reduced.
>
> In short, management essentially consists of the preparation, making, and execution of chains of decisions by regulating subsystems on the basis of information that reflects the state of the system and of its environment, as well as the intermediate and final outcomes of a system's functioning and of information concerning the fulfillment (or nonfulfillment) of decisions.[49]

Soviet theorists contend that the modification of operational decisions on the basis of new socioeconomic information or technological innovations should be viewed in a positive light, not as an indication of failure or as a cause for reproach. Soviet analysts imply that "organizational learning" is increasing in the party and state, benefitting the policy process and substantive policies in both the short and long run. Also, CPSU spokespersons acknowledge that effective decision making under dynamic, complex, and uncertain conditions necessitates the ongoing reassessment of aims, priorities, and resources, as well as a considerable amount of prudent risk taking. Iu. A. Krasin observes, "It is impossible, in politics, first to make all the necessary calculations, down to the final details, and then to take action. *Practical action is an essential component in the search for and the implementation of new possibilities*."[50] Furthermore, some commentators stress that modern decision-making and administrative "mechanisms" are needed to benefit from and contribute to the NTR. K. Varlamov notes, "In present-day conditions, the Communist Party is carrying out a whole range of measures to master the newest methods of management, which to a certain extent requires breaking down stereotyped and outdated methods of thinking and acting."[51]

Significantly, Afanas'ev, Burlatskii, and others[52] have outlined the major stages of a decision-making process, and they have stressed the need for different kinds of information at every phase, including the adjustment of previously made decisions. Afanas'ev's stages are:

1 determination of goals;
2 determination of the quantity and quality of information needed to make the decision;
3 collection and processing (systematization) of information;
4 creation of information models;
5 determination of alternative methods of fulfilling the task, attaining the goal;
6 selection of the criteria for evaluating alternatives;
7 evaluation of alternatives;
8 selection of one of the alternatives and the making of a decision;
9 adjustment [*korregirovanie*] of a decision in the course of its implementation.[53]

Afanas'ev, however, takes a flexible approach to the study of decision making. Acknowledging that respected Soviet and Western specialists disagree about the types and number of decision-making stages, he insists only that decision makers gather and process information about their system (broadly or narrowly defined) and its environments. Furthermore, he stresses that orderly decision-making theories often differ considerably from disorderly decision-making practices because of insufficient information, perceived pressures to make decisions quickly, and reactive rather than purposeful behavior.[54]

We quote below a comprehensive American critique of management theory and practice in the United States, which was reprinted almost verbatim in a Soviet journal. The Soviet authors, Afanas'ev and Mil'ner, are especially knowledgeable about and eager to reduce comparable discrepancies between theory and practice in the USSR.

What Management Theory Says Managers Do

Planning

Develop clearly determined, understood, and stated objectives.

Manage by objectives.

Utilize quantitative techniques to facilitate the achievement of quality decisions.

Engage in long-range planning to anticipate changes in the environment through forecasting.

Develop effective strategies in response to forecasts.

Organizing

Alter the job to fit the man.

Directing

Use intrinsic characteristics in jobs to motivate workers.

Alter leadership style to the situation.

Controlling

Evaluate performance at strategic points within the organization.

Correct all significant deviations between desired and actual performance.

What Managers Actually Do

Planning

Operate with vague and even contradictory objectives.

Infrequently manage by objectives.

Infrequently use quantitative techniques in decision making.

Plan in the short run.

Develop strategies in response to changes in the environment.

Organizing

Alter the man to fit the job.

Directing

Rely on money to motivate workers.

Allocate rewards as much for effort, education, age, and seniority, as for performance.

Select a leadership style compatible with the top man.

Controlling

Evaluate those activities that are easily measurable and highly visible.

Correct those deviations that are in the best interest of the manager to correct.[55]

One can only speculate about Afanas'ev's and Mil'ner's intentions in disseminating these American research findings. But a chief purpose of Soviet investigations of Western administrative behavior has been to alert Soviet managers to what the CPSU leadership perceives to be problems and opportunities as well as to borrow and adapt selected Western methods and technologies to meet these challenges. Notwithstanding the considerable differences between socialism and capitalism, the American ideas and experience described above have much in common with Soviet ideas and experience.

Burlatskii gives specific examples of what should *not* take place but often does in the Soviet management process.

> There are a number of typical causes of mistakes in the solution of problems: an incorrectly defined functional purpose, lack of a satisfactory set of alternatives, inadequate calculation of the future expenditures connected with the decision adopted (in practice, this latter is the most frequent case). Moreover, the decision itself must allow for its own modification and, if necessary, alteration under the force of changed circumstances (such as a new invention). And this is far from universally the case in practice.[56]

Burlatskii emphasizes the dilemmas of decision making. He views the determination of goals and their interconnections as "the most difficult stage in decision making" and suggests that some goals are not implemented because of "the inadequacy of the very approach to formulating a specific social task." Most problems are "weakly structured" or "not structured at all," and managers confront them with a potpourri of knowledge and intuition that is much less "scientific" than "heuristic"—the latter being "the oldest and probably still the most frequently employed method for solving economic and social problems." Noting the difficulty of obtaining pertinent information for decision making and implementation, Burlatskii decries "the inadequacy of economic and engineering criteria of production efficiency."[57]

Burlatskii underscores the potentials as well as the pitfalls of political leadership in the era of the NTR. "Anarchistic and bureaucratic tendencies" persist under developed socialism. The former "are quite dangerous for socialist society," and the latter "fetter the initiative of the masses and lead to a gulf between officials and the people." Only the directing and coordinating activities of the CPSU can counter these disruptive tendencies and can ensure the maturation of the polity and society. Emphasizing that the CPSU's personnel and operations are being "transformed," Burlatskii urges party leaders to focus on: "(1) development of scientifically grounded policy and organizational work to implement policy; (2) promotion and training of cadres; (3) determination of scientific principles and methods of administration, with extensive independence of all links of the administrative apparatus and initiative by the masses; (4) thorough supervisory control."[58]

Some of Burlatskii's observations and criticisms have been echoed by top CPSU officials. For example, Chernenko observed in 1981, "As socialism matures, the party and its bodies concentrate their attention increasingly on the formulation [*razrabotka*] and adjustment [*korrektirovka*] of the strategy of social development itself, setting tasks whose fulfillment state bodies must concern themselves with and verifying the fulfillment of the decisions adopted."[59] According to I. V. Kapitonov, head of the Central Committee's Department of Organizational-Party Work throughout the Brezhnev period, there is "much that can and should be improved" in disseminating information within the party. "We have to enhance significantly the quality of the information coming from below, from the party organizations. It is necessary to spotlight in good time new tendencies in the development of the party and society, truthfully to reflect the phenomena of life and the mood of the masses, to warn of shortcomings, and to pose questions that need to be resolved."[60]

What is striking about Soviet thinking in the 1970s and 1980s is its sensitivity to the *limitations*, both quantitative and qualitative, of even the best policy-relevant data. Afanas'ev and Burlatskii, for example, became increasingly aware of political leaders' inability to measure social policy outcomes and interrelationships; to calculate noneconomic (or economic versus noneconomic) costs, benefits, and trade-offs; and to forecast the consequences of alternative courses of action. A growing number of Soviet analysts emphasized that most political-administrative and socioeconomic decisions are made under conditions of considerable "uncertainty," "risk," "stress," "conflicting interests," and "competing priorities" (all Soviet terms).

Soviet theorists since Lenin have affirmed that political leadership is both "a science and an art." Brezhnev declared in 1970 that the political system must maximize "society's possibilities for harmonious and even development" and that the party's activities must be based on "*the facts of science and wealth of practical experience*."[61] Soviet analysts, especially in the first half of the Brezhnev administration, emphasized the need for better data from the social and natural *sciences* in order to improve the substance and outcomes of decisions and the decision-making and administrative processes themselves. But Soviet commentators, especially in the second half of the Brezhnev administration, lauded the *art* of interpreting data and of acting judiciously and effectively without adequate data. Under Andropov and Chernenko, party spokespersons praised both the science and art of leadership, placing even greater emphasis on public-opinion polling and on citizens' letters to party organizations and newspapers.

E. P. Kireev argues that the "scientific" nature of politics lies in the CPSU's will and capability "critically to evaluate the results achieved and, if necessary, to correct previously made decisions."[62] Kireev stresses the experimental or trial-by-error nature of policy making and the importance of "political feel" [*chut'e*] or "political intuition" [*intuitsiia*]. Although party policy rests on "scientific foundations," "mistakes" will be made in the formulation and implementation of policies during the NTR. Kireev suggests that mistakes can be *expected* because of the complexity, uncertainty, and dynamism of domestic and international conditions; the difficulties of predicting events, trends, and interconnections among variables; and the *human* failings of party and government leaders. He buttresses this last contention with a seldom cited statement by Lenin: "The leaders [*vozhdi*] of the workers are not angels, saints, or heroes but people like all of us. They make mistakes. The party corrects them."[63]

A. E. Bovin elaborates: "It would be naive to suppose that a knowledge of scientific politics and a correct theoretical approach to the nature and methods of political activity can serve as a *magic talisman*, pointing the way to a promised land where political reason reigns and political mistakes can be found only in history textbooks. . . . *Politics, even scientific politics, has always been and remains primarily an art*."[64] Bovin and Kireev do not belittle the use of pertinent data from the natural and social sciences in policy making and administration, but they emphasize that empirically grounded research "plays a major role in forming a scientific Marxist-Leninist world outlook and has ideological aspects."[65] In other words, one's identification and assessment of "objective" circumstances are highly "subjective" acts. Objective conditions are "mediated" through subjective perspectives, and the realities perceived and the responses produced vary accordingly. Hence, Kireev urges Soviet policy makers to obtain accurate and up-to-date scientific-technological and socioeconomic data and to study closely the political mood, needs, and capabilities of the USSR and other nations. He underscores that "the entire art of management and politics consists of determining in good time where to concentrate one's chief resources and attention. . . . Management is not a technical concept but a political process. . . . Centralized leadership does not exclude but presupposes initiative and creativity."[66]

Soviet commentators, then, are sensitive to the rational and intuitive dimensions of decision making.[67] Bovin declares:

> Politics deals with a practically infinite variety of facts and events, with highly changeable and contradictory processes and with the clash of different interests and wills. The tremendous difficulty of grasping all the links and the stormy dynamism of social relations creates a certain instability of the ground on which the politician has to construct all his calculations. . . . There is always tremendous room for assumptions and probability appraisals, so that it is important to have the ability intuitively to grasp the overall direction processes are taking.[68]

This statement encapsulates the Brezhnev administration's view of scientific leadership.

Implementation

Soviet thinking about scientific management focuses on *methods*, not on *purposes*. A central theme is that scientific-technological and socioeconomic information are of rapidly growing importance in implementing goals and policies that have already been determined by scientific leadership. What is called for is feedback that helps to fulfill rather than change targets. Increasing the flow of pertinent data to key decision-making points enables officials to modify administrative work—and occasionally operational decisions—to unexpected difficulties and opportunities. The late V. M. Glushkov, a leading cyberneticist, asserted, "Problems in managing the economy can be divided into two major classes: problems of the ideological plan and those of the technological plan."[69] He clearly implied that "technological management" decisions express the values and priorities set in the authoritative "ideological" (i.e., political) sphere. Hence, Soviet perspectives on the NUO are highly *instrumental*. They are directed to questions of political and organizational effectiveness and efficiency, not to the reappraisal, readjustment, or rejection of political and organizational goals. If anything, Soviet commentators attempt to legitimize the substance of policies by demonstrating that they are formulated, refined, and administered by "scientific" *means*.

The NUO literature contains countless discussions of "rational," "comprehensive," and "optimal" planning, decision making, and implementation. Soviet spokespersons acknowledge the persistence of many familiar problems, such as "barriers" to organizational communication and coordination, slow development of science-based high technology, and inadequate incentives for the use of technical breakthroughs in production. But Soviet observers recognize that the NTR is producing *new* problems that may become more serious as the NTR unfolds. Akhiezer notes the emergence of "nonantagonistic contradictions between technologies and organizations, between the level of production and professional skills, and between the complexity of the system to be managed and the capacity of the management mechanism, etc."[70] Mil'ner declares, "What we are talking about is an objectively formed contradiction between the specialization and the integration of managerial functions, between the division of labor and its cooperative organization, and between the administrative isolation of branch agencies and the coordination of their actions. Overcoming this contradiction is the key question in the improvement of management."[71]

Because of the NTR's complexities, uncertainties, and dynamism, the interdependence of goals and means, and the intersectoral nature of virtually all major opportunities and problems, Soviet leaders and theorists under Brezhnev were attracted to "systems" and "cybernetic" approaches to problem solving. These comprehensive and integrated approaches gave promise of ensuring that the national interest would prevail over myriad particularistic interests in the formulation and implementation of policies. "Systems analysis plays a large role in decision making," leading officials were taught in the newly formed schools of "management science."[72] With or without mathematical methods and computers, "the logical approach" of systems analysis was highly recommended, especially to facilitate "the structuring of a problem" and to help resolve well-defined problems through "operations research" and/or "heuristic methods." The key concept of "management" was defined in cybernetic terms, and the definition of "a management decision" employed the systems concept.[73] Systems approaches were to spur the transition from "extensive" to "intensive" forms of economic development and to assist party, state, and production executives in carrying out national programs.

The Management of Complex Systems. There is no unified or widely accepted systems theory in the USSR or elsewhere. But Soviet interest in systems approaches has intensified, and Soviet thinking about managing complex systems has matured considerably. From the mid-1960s Brezhnev's collective leadership was keenly interested in improving the quality and quantity of information used in societal guidance. This interest was increasingly tempered by the unsettling political implications of the feedback principle and by the theoretical ambiguities of Soviet efforts to combine Western cybernetics with dialectical and historical materialism.

Soviet enthusiasm for cybernetics began under Khrushchev in the late 1950s and lasted well into the Brezhnev period. Attacks on cybernetics mounted in the early 1970s. By the end of the decade the Soviet emphasis had shifted from cybernetics to systems analysis, from the static and closed characteristics of systems to their developmental and open characteristics, and from their informational and self-stabilizing capabilities to their problem-defining and problem-solving capabilities. Western structural-functional approaches were sharply criticized for their ahistoric and homeostatic characteristics as well as for their avoidance of both causal and normative analysis. Soviet scholars distinguished among systems operating in objective reality, systems of concepts, systems of machines, and systems including people and machines. A major Soviet institute was established in 1976 to further the theoretical development and practical applications of systems approaches, and its director, Gvishiani, emphasized the purposefulness and dynamism of systems and the need for "rational management" of systems from within and without.

> Systems analysis is concerned not so much with the study of an object as such as with the study of the *problem situation* to which it is linked, that is, ultimately, with ensuring effective *interaction* with that object. . . .
>
> As it is oriented to subordinating the study of the object to the goals of the proposed interaction, *systems analysis sheds new light on the principle of the unity of theory and practice*, which is central to Marxist-Leninist philosophy. Quite consciously, systems analysis abandons the study of objects per se, studying them only to the extent that this is needed for attaining definite practical goals. This is why systems analysis essentially stems from the urgent need to study and build up systems for which *there is not yet any mature theory* on which one must and might rely. . . .
>
> [It is important] to make a delimitation between so-called *goal-directed* [tselenapravlennye] (*adapting or learning*) *systems* (in the sense in which this term is used in cybernetics) and *goal-setting* [tselepolagauiushchie] *systems*. In the first instance, the structuring of the external interaction (the wider system) is dictated from the *outside*; in the second, it is done *within* the system being considered.[74]

Let us briefly trace the evolution of such ideas in the Brezhnev period. According to Soviet theorists in the late 1960s, a developed socialist society becomes more and more capable of guiding its own evolution in a progressive direction. Because mature socialism was perceived to be characterized by a high degree of institutional and functional interdependence, society was viewed as a "system" whose "integrative qualities" could not be reduced to the sum of its component parts. These components and their relationships were to adjust to the unusually rapid changes associated with the NTR. Although the short- and long-term goals of the system were clearly and authoritatively established, the system itself was to develop and become more complex in response to self-regulating or goal-seeking feedback from its subsystems and from its domestic and international environments. Successful management was thought to be possible only when the controlling elements of the system were informed about the effects of their actions, knew whether their purposes had been achieved, and possessed the capabilities to make timely modifications of their policies and policy-making procedures in order to reach predetermined targets.[75]

Soviet theorists viewed the study of "organic wholes" and "objects with complex organizations ('systems')" as a vital part of the scientific character of Marxist-Leninist philosophy and methodology and as "a major trend" in tsarist Russian and in traditional Soviet science.[76]

> Dialectics has shown that true, concrete thought thinks in terms of contradictions that grasp the opposing sides of phenomena in their unity. It is capable of seeing not just one aspect of a contradiction and registering it in a rigid, static concept, but all aspects of contradiction, and not only their arrangement, but their connection, their interpenetration. This means that concepts must be as dialectical, that is, as mobile, flexible, plastic, interconnected, and interpenetrating as the objects which they reflect.[77]

During the first half of the Brezhnev administration, Soviet analysts stressed that systems theory was a key element of dialectical thought. Emphasizing the evolutionary nature of change in mature socialism, they used a systems approach to underscore the importance of centrally prescribed goals and of local initiative in fulfilling these goals. Society was conceptualized as a set of interrelated subsystems, including an integrating and coordinating center that pursues overarching systemic goals. The function of a given component was determined by its linkages with other components, and all subsystems were to be dependent on the center for their linkages with the system as a whole. For example, bureaucratic and social organizations were viewed as networks that provide the directing bodies with primary and feedback information. Also, a systems orientation strengthened the Soviet leaders' claim that the USSR is becoming an all-inclusive and harmonious social order. They acknowledged that Soviet society is differentiated, even stratified, but they insisted that social cohesiveness can and must be enhanced by the dynamic interaction between the subject and objects of management. Hence, the center's crucial functions were to preserve systemic unity under changing internal and external conditions; to energize the system and guide it to higher levels of purposeful and organized activity; and to protect the system against enervation, aimlessness, and disorganization.

Throughout Brezhnev's collective leadership, and especially up to the mid-1970s, Soviet theorists emphasized that the generation, control, and use of information are critical to the functioning and evolution of all systems. Afanas'ev, in the major Soviet work on information and management, defined information as "*knowledge that is utilized for the orientation of systems, their active operation, and management, i.e., in order to maintain their qualitatively distinct properties and to improve and develop them*."[78] Elsewhere he elaborated:

> The management of any dynamic system is organically connected with information, the flow of information processes. The constant circulation of information among components of the system and between the system as a whole and its environment is an essential attribute of management. Information processes make it possible for the system to interact purposefully with surrounding conditions, to coordinate and subordinate the relations of its own components, to guide their movements and its own movements toward the previously programmed goals. The result is that the self-managed system can preserve its integrity, its specific qualities, and in many cases (biological and social systems, for example) not only preserve but perfect and develop itself.[79]

Most important, Afanas'ev simultaneously *emphasizes* the need for goal-directed feedback and *deemphasizes* goal-setting feedback. To be sure, Afanas'ev encouraged fellow members of the Brezhnev leadership to seek more and better information in order to formulate, implement, monitor, and adjust policies. But he insisted that top CPSU officials control the parameters of political choice and use feedback from the specialized elites and masses only if it were communicated through officially approved channels, if it enhanced the Politburo's and Secretariat's power and authority to make decisions of major importance, and if it helped the national leadership to design and execute feasible social and economic programs. On the one hand, feedback was to spur socioeconomic progress and occasionally to modify the party leaders' conceptualization of progress. On the other hand, feedback was not to constrain, circumscribe, divide, or delay. The Soviet dissident ideas that policy should emerge from a "dialogue" between the center and periphery, between the political majority and minority, and between the controlling minority and controlled majority are anathema.[80] Afanas'ev concludes that *responsible* guidance is not necessarily *responsive* guidance under developed socialism. Managerial rationality and accomplishments depend largely on information that helps the CPSU leadership to preserve the essentials of the present-day polity in rapidly changing domestic and international environments.

Systems theory calls for a comprehensive and integrated approach to the management of a mature socialist society. Soviet analysts have long praised central planning and management, and they now view systems ideas as means of improving these capabilities. Although Soviet spokespersons used the phrase "problem situation" more and more after the mid-1960s, it was closely linked to systems analysis only in the mid-1970s. Soviet commentators stressed that systems approaches could help to define and resolve problems, especially by revealing their structural and behavioral interconnections with other problems, opportunities, and circumstances. Systems theory was expected to clarify societal and organizational goals and the tasks necessary to fulfill these goals. I. V. Blauberg, V. N. Sadovskii, and E. G. Iudin emphasize that many complex and interrelated problems "are not solved at all or are solved incorrectly or ineffectively only because they have been wrongly formulated, because not all the essential aspects of the problem or constraints imposed on its solution have been taken into account."[81] Furthermore, they argue that systems analysis can improve the policy-making process as well as policy outcomes. "Of prime importance here is the structuring of problems—determining the real goals of the system and the alternative ways of achieving them, together with the analysis of external conditions and constraints."[82]

Summarizing contemporary Soviet perspectives on the application of systems approaches to management problems, Blauberg, Sadovskii, and Iudin affirm:

> The strong emphasis on the practical orientation of systems ought not to conceal its profound methodological content. In its most simple and general form, systems analysis may indeed be reduced to the rationalization of intuition in managerial activity and to the search for rational ways and means of simplifying complex problems. That is where the secret of its success and rapidly growing popularity lies.

> The theoretical and methodological significance of systems analysis, however, goes far beyond these utilitarian bounds; it is determined by the fact that systems analysis is perhaps the most serious of all attempts that have been made to construct and realize a methodology specifically adapted for solving systems problems that are becoming ever more frequent and acute in modern science and practice.[83]

The authors admitted in 1977 that the systems movement was "an assemblage of nontrivial problems" rather than "a systematized summary of results achieved." They also acknowledged the conceptual, methodological, and philosophical ambiguities and inconsistencies of systems approaches at home and abroad. But they noted with satisfaction that "the cybernetic boom has passed its peak," and that the "unjustified sensationalism" associated with systems ideas was undergoing a "similar evolution." They insisted that "all complex control systems and technical modelling of intellectual activity" would be "inconceivable" without systems analysis and that major (especially global) economic and social problems would be "impossible" to study and solve without systems techniques.[84]

There are diverse Soviet interpretations of "systems" goals, problems, structures, interrelationships, strategies, responses, and so on. But common to virtually all Soviet systems ideas is an *instrumental* orientation. Systems theory is viewed as a means of clarifying, organizing, and pursuing values, while not embodying, assessing, or choosing among them. Although this analytical technique is claimed to be value-free, the analyst is to use it to further "progressive" values. Such values are not inherent in any particular type of systems approach because the rationality of a system's goals is subjectively determined by its managers. As Gvishiani puts it, "the most important practical recommendation of systems analysis" is to begin with a careful, comprehensive, and correct formulation of one's objective as well as an understanding of the problem to be solved and the methods of solving it, all of which "reflect the interests and views" of a specific class or group of people.[85] Systems approaches, in the service of the CPSU leadership and conjoined with its Marxist-Leninist world outlook, are "a powerful instrument of cognizing and transforming society along progressive, socialist, and communist lines."[86]

Soviet systems theorists have established hierarchies of goals such as "trees," "ladders," and "chains." Communism, socioeconomic progress, and the well-being of the Soviet population are the most frequently cited "strategic" goals, and "tactical" subgoals are countless. Soviet commentators are sensitive to trade-offs among societal goals, but they have little to say about how such choices are being or should be made. Afanas'ev maintains that systems approaches are helpful in ascertaining whether goals are "coordinated," "neutral," "contradictory," or "incompatible."[87] However, the establishment of priorities and ends-means linkages, as well as the assessment of the consequences of these priorities and linkages, are value-laden. Afanas'ev sidesteps the difficulties of choosing among vague and conflicting purposes and of evaluating political outcomes. Instead, he underscores the importance of communication and control mechanisms and the difficulties of conceptualizing interconnected problems and problem-solving approaches. He concludes: "Systems analysis is essentially a method for bringing order into a problem through elucidating a system's objectives; constructing a tree of objectives, alternative ways of achieving them, and elements of interdependence within the system in the course of carrying out these alternatives; listing the problems to be solved; and identifying constraints and conditions within the system and its environment."[88]

As the 1970s progressed, Soviet analysts placed more and more emphasis on the open and dynamic nature of social systems as well as on the powerful reciprocal influences of systems and their environments. This shift is not surprising, given the Soviet polity's increasing permeability to influences from Soviet society (especially from the scientific and technical intelligentsia) and the Soviet economy's increasing permeability to influences from the world economy (especially from changing prices and supply and demand). Soviet commentators were well aware of these developments but usually highlighted the reverse relationship—namely, the growing capabilities of the USSR to influence global affairs and of the Soviet polity to influence Soviet society. Afanas'ev explains:

> *The higher the organization of a system the greater its sensitivity to its environment, on the one hand, and the more active its influence on the environment, on the other.* Interacting with its environment, an inanimate, inorganic system is destroyed, absorbed by the environment. A living organism adapts itself to its environment. Man and a social system not only preserve their integrity in conditions of the changing environment but *change their natural environment*, transform it according to their interests and needs.
>
> The environment is an important factor of differentiation and integration of social systems. A system derives material from the environment to replenish and renew its components, to *improve its structure*. Integrating the elements of the environment, a system trans-

> forms them *according to its own nature*. The environment, while exerting a continuing perturbing influence on the system, forces it to restructure itself, to neutralize or to assimilate this influence.[89]

Hence, party spokespersons see the structure of the Soviet system as evolving but its essence as immutable, even in the era of the NTR and in a rapidly changing international system.

During the late 1970s, however, Soviet theorists increasingly linked the concepts of "goal" and "end-result," and they placed greater emphasis on bridging the gap between aspirations and accomplishments. Soviet analysts were keenly aware that bureaucratic units define problems and set goals that often obstruct, even subvert, the public interest as defined by top party bodies. CPSU leaders and commentators perceived a need for modern goal-oriented approaches, such as target programming, and distinguished sharply between the immutable goals of the social system and the improvable programs and performance of the managerial system. According to Afanas'ev, administrative behavior should be brought into accord with centrally determined aims that must not be scaled down or compromised in acquiescence to organizational pathologies.[90]

Soviet commentators have expressed a wide variety of views about how to curb particularistic interests, especially in the management of the economy. Computer specialist Glushkov, with barely concealed impatience, criticized officials who make a "fetish" of all aspects of the present-day economic system and who "naively" believe that it possesses "magical" and "unlimited potential."[91] He affirms that improvements in the "economic mechanism," organizational structures and behavior, and management methods and technology—especially information processing software and hardware—are closely interrelated and that they must proceed hand in hand or not proceed at all. Glushkov declares, "*No system of planning indicators, prices, and criteria for evaluating the performance of different kinds of workers will function successfully if there is not created at the same time a developing (adequate) accounting-control (information) system, which binds all levels and links of the economy into a unified complex*."[92] Needless to say, the establishment of such a communication and control system would necessitate dramatic changes in the relationships among present-day Soviet institutions and in their processing and use of information.[93]

Mil'ner, calling for moderate change, distinguishes between "deductive" and "inductive" approaches to the restructuring of organizations. The former approach is to be used for all major decisions, the latter for details only. Feedback is to be primarily goal seeking rather than goal setting. Target programming must further both democracy and centralism and must strengthen the existing branch and territorial institutions. Target programming, Mil'ner claims, can increase the center's capacity to subdue parochial interests and to direct regional initiatives toward the fulfillment of the national interest. The chief short-term objectives are to combat "departmentalism" and "localism," to use all resources more effectively and efficiently, and to improve economic performance. The chief long-term objectives are to heighten "purposeful, flexible, and successful management in all links" as well as to expand rights, opportunities, and responsibilities.[94] Mil'ner is well aware that new and dynamic organizational methods, such as target programming, will necessitate the "speedy and well-grounded restructuring of management organizations, in conjunction with changes in the national economy." He cautions that organizational innovations should be implemented with great care and only after "resolute efforts" have been made to eliminate existing shortcomings and to mobilize the "vast unused reserves" of the current branch and territorial institutions. Although this formulation could justify endless delay or inaction, Mil'ner advocates "organizational experimentation" and changes in "some of the structures of management."[95]

Afanas'ev, taking a particularly conservative stance in the early 1980s, observes that target programming is not a "panacea" and warns of the *negative* consequences of administrative reforms. All organizational restructuring affects the "fragile fate of many people" and has "psychological costs."[96] Alluding to Stalin's assault on the party and Khrushchev's assault on the ministries, Afanas'ev ominously declares that "groundless reorganizations" cannot be justified in the name of "high-minded goals" or by means of skillful "packaging." Before altering institutional relationships it is necessary to be "absolutely certain" that the existing relationships are incapable of resolving the problems at hand.[97] It is, of course, impossible to be "absolutely certain" about such highly politicized judgments whose consequences for different segments of the population cannot be known for some time. Also, Afanas'ev does not specify *who* must be without doubt, let alone *why* or *when*. This amounts to a thinly veiled warning that conservative CPSU leaders will engage in more than the usual second-guessing and recriminations if major organizational reforms are launched but fail to produce unspecified benefits quickly. Afanas'ev implies that the advantages of preserving key traditional administrative practices outweigh the risks of moderately altering these practices. Innovative approaches, especially multifaceted, comprehensive, and potentially disruptive ideas, such as target programming, must be tailored with the utmost care to present-day institutions, not vice versa.

Why does Afanas'ev conclude a discussion of target programming by calling for "circumspection" and by cautioning against "the loss of realism for even a minute"? Why is Afanas'ev so concerned about the costs rather than the benefits of target programming and especially of institutional changes? What are these costs and benefits, and who will suffer and gain? The beginnings of an answer lie in Afanas'ev's observation:

> The extreme importance and the urgent need for a comprehensive and systems approach in mature socialist society can be explained primarily by the fact that this society is characterized by unprecedented integration, interconnectedness, organic interdependence, and reciprocal influences in the organization of its public life and in its economic, social-political, and spiritual processes. *Purely economic, social, organizational, or any other kinds of problems do not and cannot exist here; these problems and elements are interdependent and influence one another.*[98]

In a word, political and economic reform are thought to be one and the same.

The difficulties of managing complex systems in the USSR are enormous, and Soviet leaders and theorists recognize this fact. But Afanas'ev, describing desired rather than actual conditions, implies that Soviet society *is* well integrated, interconnected, and interdependent. On the contrary, the fragmentation of the Soviet bureaucracies poses a formidable challenge to the theorists and practitioners of systems approaches. Virtually all officials and analysts have strong departmental or local interests. Also, most have a vested interest in incremental policy changes, and some seek reformist and reactionary changes. Hence, for both institutional and policy reasons, they perceive different problems and prescribe different solutions. The management of research and development is an important case in point.

The Management of Research and Development. The *goals* of research and development in the USSR are a key dimension of "scientific management," according to Soviet theorists. The rationalization of R and D institutions and the effectiveness of science policy are deemed critical to progress in all spheres. Analyzing the relationship between science and organization, Iu. Sheinin contends:

> The USSR is the first state which has consciously brought about a conjunction of science and organization. This it has done at every

> stage of its development. The agencies by which the state directs science have worked out and implemented in a balanced manner, with the broad support of scientists, technical workers, and all the other working people, a coherent system of organization and direction of scientific activity. This system is an organic part of the metasystem of economic planning, constituting one of its ever more important and leading elements. . . .
>
> While the influence of science on society's development tends to increase as it becomes ever more organized, the influence of organization on society is increased as it acquires scientific character. As a result of the growing interpenetration of science and organization in socialist society, the two tend to coalesce into scientific organization in the broad and narrow senses of the word: the scientific organization of society and the scientific organization of science itself.[99]

Especially in the era of the NTR, Sheinin argues, "socially organized science and the scientific organization of social production" must be closely integrated because they have "the leading role to play in remodelling social conditions in accordance with the requirements of scientific reason in the sphere of the productive forces." Key "investments in the future" include scientific education and improved scientific information systems.[100]

The scientific management of science is rooted, first of all, in the complex structural development of pure and applied research. The increasing differentiation of scientific activity requires purposeful reintegration. Unguided and fragmented R and D are wasteful. As L. K. Naumenko notes, the relationship between the increasing specialization of science and the need for integration within and among the natural and social sciences is a basic contradiction that has to be managed. "The creative potential of science grows only when the reciprocal influences among the sciences and comprehensive scientific research are not based on a mechanical fusing of the findings of various specialized kinds of research. Such potential grows only when the return from every link of the system is increased, when research in every field is based on the total potential of all sciences and enhances it. Creativity is a collective process."[101] Hence, the integration of creative ideas, as well as their creative application, are primary goals of Soviet sociologists of science.

The scientific management of science is also rooted in the shift from the extensive (quantitative) to the intensive (qualitative) phase of development. Afanas'ev maintains, "The possibilities of extensive scientific development are limited. Science cannot keep on increasing its share of the national income or its research force indefinitely. Hence, the urgent need for an intensification of science, that is, for economizing of the time spent on research and development, and for making the work of researchers and research teams more effective."[102] Among the dimensions of the shift from the extensive development of science are "improved methods and means of research, further specialization and cooperation in the work of researchers and research teams, improvement of their equipment, their experimental base, gradual mechanization and automation of research, improved organization and management of scientific work, and the strengthening and developing of communication between science and production and between science and all of social life."[103]

The scientific management of science is rooted, finally, in the historical challenge of the transition from socialism to communism. According to Afanas'ev, R and D administrators are to "create conditions in the scientific collective and science as a whole in which the knowledge and experience of every individual scientist can be used to maximum effect so that every scientist capable of opening up new paths of progress for science, technology, and management . . . should have the possibility of doing creative work, of producing new ideas and putting them into effect, and of being suitably rewarded in both material and moral terms."[104] The microcosm of scientifically managed science will be used as a model for the management of the whole society. Afanas'ev contends that "the basic objective of the management of science tends to coincide with the basic objective of the management of the developed communist society of the future . . . the fullest possible all-round application of each individual's creative capacities in order that society's capacities may grow, and also the fullest possible use of society's capacities for the all-round growth and development of every individual."[105] Hence, the scientific management of science is to stimulate the creativity of scientists and to develop structures and incentives enabling scientists to satisfy pressing social needs.[106]

The science of science [*naukovedenie*] is comprehensive in scope and includes the following fields: the general theory of science; history of science; planning and management of scientific research; theory of scientific forecasting; technological development of the operational base of science; mathematical modelling of scientific activity; scientific organization of labor in science; psychology of science; ethics of scientific activity; aesthetics of scientific activity; science and law; language of science; and others.[107]

This is not merely an extensive list. Instead of juxtaposing various scientific disciplines, Soviet analysts view the management of R and D in systemic terms. They hope to create a comprehensive and synthetic science of science. As S. R. Mikulinskii and N. I. Rodnyi argue, "The science of scientific organization emerges from the study of the history of the development of science and technology, the sociological problems of science, and the psychology of scientific-technological creativity. The synthesis of these studies enables one to develop the theoretical basis for the rational organization of scientific research and the planning and management of science."[108]

Soviet theorists aver that the scientific management of science rests on a judicious combination of political skills, organizational capabilities, and scientific knowledge. The science of science is to provide decision makers with theoretical and practical knowledge, especially for science and technology policies. Knowledge of scientific findings and skillful administration of scientific institutions are deemed essential to the planning and management of socioeconomic behavior in the short and the long run. Soviet analysts maintain that innovative R and D must be stimulated and promptly incorporated into policy analysis and forecasting. Under advanced capitalism a sociology of science has emerged, but its nature and uses have been allegedly distorted by the fragmentation of political power and by the dominant influence of the profit motive in R and D. Under mature socialism, however, the science of science is thought to be a critical component in the acceleration of scientific, technological, socioeconomic, and political-administrative development.

Soviet observers argue that science in the age of the NTR can and must be a highly organized activity. The science of science is an emerging discipline that investigates scientists' behavior in collective settings and recommends ways to organize scientists for maximum productivity. Soviet commentators assert that the organization of a laboratory group, sector, or institute, as well as the style and level of management, the interpersonal relationships, and the creative atmosphere in the collective, are matters to be studied by sociologists and psychologists of science. For scientific collectives to be more effective and efficient, an appropriate social environment inside the collective must be established and maintained. Psychologists, social psychologists, and management specialists are to spur scientific creativity by linking the individual and the collective.[109] According to V. A. Frolov, individual creativity is transformed into a socially useful activity primarily in an organized setting and through the "spirit of the scientific collective."[110] Scientific freedom is the right to participate in research teams that strive to discover the laws of the natural and social sciences and their "progressive" practical applications.

Creativity is viewed as a collective process on at least two levels.

Societal needs increase the probability of a scientific discovery or technological innovation in a particular area, and the party-state funds research to meet these needs. Also, numerous scientists and engineers evaluate the creativity and practical consequences of a discovery or prototype. Scientific activity, then, is directly affected by the rationality of the collective mechanisms for assessing the significance of research findings and their uses. Although the individual's imagination and investigations are to be encouraged, the collective must nurture only those initiatives that serve the public interest as defined by the CPSU leadership.

Soviet analysts examine many psychological and social dimensions of collaborative research, and they see the collective as a crucial means of linking diverse fields of advanced knowledge. They investigate the structural roots of collaboration and conflict within and among scientific collectives. Also, they call for studies of communication patterns within and among scientific bodies in order to determine how researchers might perform interrelated tasks more successfully. Furthermore, they contend that the role of "science organizers" must be clarified and enhanced. Such organizers—chiefly management specialists rather than practicing scientists—are to play an important part in increasing research productivity in the era of the NTR.[111] This final assumption is the most problematical, unless administrators can provide researchers with the equipment, materials, and staff that are frequently in short supply in Soviet laboratories. The soundest assumption is that new incentives can develop interdisciplinary skills and interorganizational cooperation. Because of the high degree of competitiveness, compartmentalization, and specialization in the Soviet scientific community, more interdisciplinary and interorganizational efforts could prove fruitful.

According to Soviet commentators, the effectiveness of scientific activity can be increased by viewing science as a network of flexible and dynamic organizations. Gvishiani argues, "Science policy must change and in turn guide the development of science, as new fields of science emerge and as the social functions of science evolve."[112] The successful management of science requires an organizational strategy. On the institutional level, rigid structures must be supplemented or supplanted by mobile structures that can stimulate innovation and that can confront complex and interdisciplinary scientific-technological problems. On the behavioral level, the vested interests of existing scientific institutions pose a difficult dilemma. As Mikulinskii opines, "It is easier to create dozens of new laboratories and departments than to shut down an old worn-out unit. This leads to unproductive expenditures and lowers the effectiveness of scientific research."[113]

Organizational flexibility is sometimes linked to an "integrated program approach" as well as to contractual methods of developing closer ties among innovators, manufacturers, and consumers. Also, science centers and science-production associations are perceived to be particularly important means of developing a flexible science-technology-production cycle. The "rational siting" of regional research complexes is considered essential to the creation of new industrial capabilities such as those in Siberia and the Far East.

Criteria and standards to evaluate the results of R and D are a major concern of Soviet managers of science. Such assessments are crucial in selecting the lines of research to be funded. The allocation of scarce resources is not to be determined by the egos of ambitious scientists or by the particularistic interests of competing research institutes. Objective indicators are to be developed to evaluate the social significance of research and to rationalize managerial decisions. The contributions of experts are vital to the evaluation of specific discoveries and inventions as well as to the overall development of R and D.[114]

For example, G. M. Dobrov views scientific research as a complex system of generating, transmitting, and transforming information. He creates communication models to improve the planning and management of scientific progress and develops mathematical formulas to express various relationships between information flows and scientific innovations. Dobrov declares: "The volume, complexity, and responsibility of the management of science cannot be compared with any specific scientific task, taken separately. What is needed here above all is the synthetic processing of information on the functioning of scientific institutions, the formulation of tendencies, the comparative appraisal of levels, the preparation of variant solutions, forecasting, and the continued clarification of the hypotheses drawn from forecasts."[115]

Also, control over scientific expenditures is thought to require the development of a new subfield of economics. According to L. Gliazer, the economics of science must ascertain "the objective patterns of the production, distribution, and consumption of social resources in the sphere of scientific research and substantiate the ways of using knowledge of these patterns to obtain results that are optimal from the standpoint of the effectiveness of the national economy."[116] Gliazer argues that economic studies of scientific activity help "to determine the basic relations and dependencies of scientific production, to establish indices characterizing the most important parameters of the development of science to determine the most effective direction of research and development, and to create stimuli that accelerate scientific and technical progress." Hence, successful research helps to subordinate scientific activity to "the deliberate, purposive direction of society."[117]

The scientific management of science includes a cadres policy that, according to Gvishiani, must solve four basic personnel problems: determination of the major professional needs of cadres engaged in fundamental and applied research; establishment and maintenance of an appropriate distribution of researchers among the leading branches of science; training of adequate numbers of specialists in each scientific branch; and preservation of a "correct balance" between highly qualified scientific personnel and auxiliary scientific workers. Moreover, an effective cadres policy recruits talented youth into careers in science. As Mikulinskii observes, "The high social prestige of scientific personnel, their relatively high pay, and the attractiveness of the scientist's profession have brought about a gravitation toward science, while the critical need for scientific personnel sometimes leads to a lowering of demands in the selection process. Therefore, it has become necessary to work out methods for the early diagnosing of ability for scientific work and the selection of cadres for work in science."[118] Furthermore, Mikulinskii insists that only those people capable of working in the collective setting of modern science be recruited. But he portentously adds, "This makes necessary a thoroughgoing and complex restructuring of the psychology of scientific personnel."[119]

Soviet policy makers have asked scientists to help them manage the development and utilization of science. Some scientists have complied more enthusiastically than others, and all have found it difficult to develop "social indicators" that objectively ascertain the effectiveness of science and technology. Gvishiani contends that the centralized guidance of R and D is essential for the following reasons: the complexities of modern science; the high cost of the technological infrastructure for science; the long-term planning necessary to balance theoretical and applied research; the need to coordinate vastly increased information flows; the benefits of exchanges with scientists from other countries; and the protection of scientists' and engineers' patent rights.[120]

Although the centralized management of science is not to be arbitrary or subjective, strong leadership may be bureaucratic, and innovation may suffer. V. Zh. Kelle and Mikulinskii state:

> Modern science is inconceivable without a modicum of centralized leadership, but excessive centralization already tends to engender the danger of bureaucratization in science. That is why there arise the

> questions of improving the management of science and working out forms ensuring managers' flexibility and capability to respond swiftly to the new requirements that come to the fore in science and production and that help to establish the conditions for the utmost development of creative initiatives within scientific collectives and on the part of every creative scientist.[121]

The challenge, Gvishiani observes, is to establish a "balanced combination of centralized management of science with the maximal development and actualization of all of the creative potentials of scientific collectives and of each scientist."[122]

The Soviet political leadership is to respect the relative autonomy of scientific institutions, according to Gvishiani. "Science as a complex social phenomenon is subject in its development to specific laws of development. Without discovering them, it is impossible correctly to organize, plan, and manage scientific activity and even to create an effective science policy."[123] Information flowing from the scientific community to the party-state is to be expanded and regularized, enhancing political intervention in R and D. For example, many Soviet scientists and most party officials support research on social indicators. This can be explained by the fact that both groups endorse or acquiesce in the use of scientific advances for purposes determined by CPSU leaders.[124] The policy-making bodies establish the "external" goals that guide the development and productive utilization of science, but these goals must conform to the "internal" needs of scientific progress. The balance between the central direction of science and its organizational autonomy is highly politicized and ever changing.

Top CPSU organs are to define the strategic goals for scientific development and to plan the evolution of the R and D institutions. Party and state leaders determine scientific-technological and socioeconomic needs and their interrelationships. They identify the lines of pure and applied research that have the greatest social significance and provide an infrastructure for research. P. A. Rachkov asserts:

> Creating and improving a new type of social organization of science, *the* CPSU *and the socialist state acquire through this process growing importance in the coordination and activization of science*, as well as increasing its influence on all aspects of social life. . . . The further improvement of the system of party and state leadership of science, its rational combination with the growth of organizational involvement in collective scientific activity, have growing significance in the utilization of the advantages of the socialist structure and in the maxi-

mum transformation of science into a direct practical force for social progress.[125]

One major reason why Soviet leaders and theorists have embraced systems approaches lies in the ambiguity surrounding the definition of "subsystem autonomy." Science is perceived to be an open subsystem functioning in a much larger social system whose political subsystem establishes goals for and integrates all subsystems.[126] However, the relationships among subsystems are not rigidly specified and are the subject of considerable dispute. The tension between autonomous scientific development and politicized scientific development persists. "Science is not only an integral element of society as a system, but it is a complex and self-regulating subsystem—a relatively independent social organism," observes A. Titmonas.[127] Significantly, the "independence" of science is increasingly defined in terms of its "interdependence" with policy institutions. Soviet analysts do not minimize the tension between R and D and political-administrative aims. Rather, they include it among the major contradictions to be rationally managed.

The *accomplishments* of R and D in the USSR, according to Soviet and Western analysts, contrast sharply with the desired conditions and theoretical concerns discussed above. No single organization directs or coordinates pure and applied scientific investigations. The chief sets of institutions conducting R and D—the Academy of Sciences and its affiliates, the universities and other advanced educational bodies, and the industrial and agricultural branch ministries—are remarkably fragmented. The need for cooperation and communication, to say nothing of a "systems approach" to or the "rational management" of Soviet science, is real and pressing. Cocks affirms:

> It is important to emphasize that the popular Western image of a tightly centralized and coordinated Soviet science and technology effort has never corresponded to reality. Central planning and management of R and D is still highly imperfect. Science and technology planning has always been much more rudimentary than economic planning. Although much more centralized and comprehensive than the American system, the Soviet approach is far from the holistic model that it is sometimes portrayed to be. The Kremlin's reach in science policy continues to exceed its grasp. Aspirations outdistance capabilities. There are still many holes in the whole. The interplay of multiple agencies with diverse perspectives, different wills, and competing interests continues to constrain the actions and to limit the

capabilities of central authorities to formulate and implement coherent policies in science and technology.[128]

Cocks stresses that two quite different "systems" of institutions have shaped Soviet R and D. The first is the Stalinist command and control system of economic planning and management, which has "a strong anti-innovative bias [and] remains fundamentally oriented to the expansion of existing patterns of production and technology." The second is the less powerful and more differentiated system of organizations that Khrushchev and Brezhnev developed in response to the NTR and that is primarily concerned with the innovative aspects of science and technology policy and performance. "Each system has its own plans, budgetary practices, incentive schemes, and integrating administrative organs. Typically, however, there is lack of coordination between the basic and supplementary systems. Indeed, they frequently work at cross-purposes to each other."[129]

Because Cocks and other Westerners have elaborated on these themes,[130] we will merely underscore that Soviet analysts began to confront R and D problems with greater frequency, frankness, and foresight during the Brezhnev years. Although Soviet criticisms of the organization and implementation of scientific activity produced relatively few changes, the substance of these criticisms and the fact that they were expressed publicly constituted progress toward the "rationalization" of R and D.

Three illustrations regarding science and technology policy will suffice. The first concerns the poor quality of Soviet information about R and D personnel. As Philip Hanson notes, "The lack of published data on the number of qualified scientists and engineers engaged in research and development (and on their branch distribution) is a notorious difficulty for Western students of Soviet science policy. What is more, a recent article in *Pravda* asserts that the data are not merely unpublished; *they are not collected at all and are unavailable even to Soviet planners*."[131] Indeed, the branch ministries have a vested interest in withholding accurate information about their human and material resources from the two major agencies engaged in interbranch coordination, Gosplan and the State Committee for Science and Technology. Without sufficient data about scientific personnel, however, the planning and management of R and D are seriously impeded.

Second, the fragmentation of R and D in the Soviet Union produces considerable "groupism" [*gruppovshchina*] and "bureaucratism" [*biurokratizm*]. These "negative phenomena" are even more detrimental to Soviet science than conflicts within scientific collectives, intellectual conformity, and unethical professional behavior, according to V. G. Kostiuk.[132] The former problems he views as "objective" contradictions, the latter as "subjective" contradictions. Kostiuk acknowledges that groupism and bureaucratism exist in many segments of society other than science, and he implies that the eradication of these insufficiently studied characteristics of the scientific community depends very much on their eradication in the larger "social system" and in its other "subsystems."

Groupism is the pursuit of particularistic interests rather than the public interest, and bureaucratism consists of such pursuits in department and local management bodies. In Soviet science, Kostiuk argues, groupism manifests itself in the composition of many scientific collectives. Members are selected not because of their professional qualifications but because of the perceived need for different forms of "protectionism, which are based on family circles [*semeistvennost'*], acquaintances, connections with 'necessary' organizations and people, and so on." Other problems include the "careerism" of leading researchers "in the field of their scientific interests and outside of these interests," as well as the formation of "schools of thought" whose members are certain of the "infallibility" of their scientific views and exclude or belittle all "dissident" professionals who do not agree with them. The "school" of T. D. Lysenko is cited as a "classic" example, but Kostiuk ominously observes that "such phenomena can also be found in the contemporary period."[133]

Groupism manifests itself in the activities of scientific collectives too. Kostiuk criticizes the personal ambitions of unnamed leading scientists and entire R and D teams that strive for special and illegal privileges in the financing of their work, the supply of materials and equipment, the assignment of cadres, and the publication of their findings. Kostiuk supports the concentration of scarce resources on high-priority projects. But he calls for the curbing of the "subjectivism" and "voluntarism" that usually result from the self-serving efforts of "strong personalities" and from "group pressure" under a wide variety of conditions (e.g., success, failure, and routine operations).[134] Kostiuk does not specify how or by whom such behavior should be curbed. He implies that the R and D "subsystem" and the broader socioeconomic management "system" must share the responsibility for the persistence of groupism and must take innovative measures to minimize it.

According to reformist Soviet commentators, the NTR *heightens* bureaucratism in many fields. V. A. Rassokhin maintains that serious contradictions are developing between "the increasing need to resolve the fundamental problems of applied science and the growing strength of the departmental orientation to the resolution of current tasks." Also, a contradiction is deepening between "*the mounting need for interbranch coordination*, especially for a unified state science and technology policy, and *the lack of interbranch (nondepartmental) scientific centers*, which could formulate and implement a unified science and technology policy among the interconnected branches of the economy."[135] Kostiuk adds that careerism is an especially insidious form of bureaucratism in Soviet science because the well-established principle of "one-man management" [*edinonachalie*] gives leading scientists considerable authority over their research teams and departments. Bureaucratism is not limited to careerism and conformism *within* scientific collectives; it is prevalent in relations *among* collectives—especially in the form of "red tape, callousness, and manipulation of the thinking and behavior of individuals and collectives; withholding of information about their activities; the making of important decisions by 'a narrow circle of responsible officials'; and provoking squabbles rather than resolving contradictions."[136]

Soviet analysts perceive such problems to be serious. They recognize the continuous need to evaluate the priorities, results, and efficiency of scientific research as well as to design, manufacture, install, and replace the technological applications of scientific advances. Kostiuk concludes, "Negative institutionalization—the manifestation of active, aggressive bureaucratization in science—is an extremely harmful and dangerous tendency in the development of science."[137] Hence, Soviet scientists and science organizers call for more flexibly structured R and D, more collaboration among researchers in different institutes, better coordination between scientists, engineers, and production executives, improved manufacturing and distribution of modern equipment and materials, and freer exchange of scientific and technical information. Scientists much more than science organizers favor closer Soviet ties with *international* leaders in science and technology.

Third, the absence of a single institution managing R and D has generated appeals to strengthen the integrative functions of the State Committee for Science and Technology. The situation during the Brezhnev administration was "paradoxical," in Rassokhin's view. "On the one hand, basic science and industry are being squeezed more and more into a departmental-branch framework, and, on the other hand, the USSR State Committee for Science and Technology has no proper scientific base—a role that can be played only by leading institutes in the technical sciences."[138] In other words, science and technology policy is uniform only in theory, and the major state agency responsible for making it uniform in practice possesses neither the political-administrative power nor the R and D resources to do so. The problem is compounded because the present fragmented "system" is firmly supported by Soviet legislation.

Rassokhin underscores the importance of the "legal monopoly" held by the industrial and agricultural agencies' research institutes; the lack of incentives for interagency cooperation in stimulating, producing, and utilizing discoveries and inventions; and the rewards for making numerous incremental adjustments in traditional technologies rather than for generating and disseminating a few technological breakthroughs. Emphasizing the role of law in establishing a science-technology-production cycle, he calls for major judicial and administrative changes.

> Basic science in the sphere of industrial production urgently needs to be independent of departmental interests and branch limitations. It is possible that without such independence the continued existence of "big science" in our industry will be jeopardized. . . .
>
> Major problems of scientific and technical progress in our economy cannot be solved through departmental-branch structures alone.
>
> Thus, there is a growing contradiction between the objective requirements of interbranch integration in the development of science, technology, and the economy, on the one hand, and the interests of departmental structures that are aimed above all at further improvement of traditional technology within a narrow branch framework, on the other. This is graphically evident in the negative consequences that result from legal monopolism by a "head institute," which is almost always a research institution belonging to the producer-branch. . . .
>
> It would seem that the time has come to redistribute rights in the sphere of scientific and technical policy. The decisive rights in this area should be assigned not to the head institutes of the producer-branches but to *independent scientific institutions capable of rising above departmental interests* and basing their actions on the objective logic of scientific and technical progress and the higher interests of the national economy. Furthermore, the right to resolve questions relating to the development and introduction of scientific and technical achievements (ones that do not have to be submitted to top-ranking academic institutes) should be granted to the head institutes of consumer-branches. This would mean that the decisions made would be more objective and would be in keeping with the interests of the state and the national economy.[139]

Rassokhin has argued since 1980 that the solution to R and D difficulties—or, at the very least, the way to build a few more links between pure and applied research—is to establish "a fourth system of scientific institutions."[140] In addition to the academies of sciences, higher educational, and branch-ministerial systems, the State Committee for Science and Technology would organize and "directly manage" a network of interbranch research institutes. Such a network would embrace many existing institutes and would operate "alongside" the present three systems. Rassokhin does not call for substantial additional capital investment in the new organizations. Instead, he advocates the legally grounded expansion of the powers of the State Committee for Science and Technology.

The Soviet response to Rassokhin's idea has been mixed.[141] In 1982 Hanson summarized the different appraisals.

> Nobody disputed the importance of departmental barriers as a problem in industrial research. When the proposal was opposed, though, it was on the grounds that it would not by itself solve the problem. It might be thought that senior officials of the State Committee for Science and Technology would be happy to extend their empire. Perhaps some of them would be, but the situation is not as simple as that. One deputy chairman of the committee recently called for the creation—preferably under the Academy of Sciences—of large suprabranch centers for research and development, with experimental production facilities.
>
> That proposal envisaged the new research and development centers as national institutions. Other recent suggestions stress the attractions of regional interbranch arrangements for research and development [and] the role that local party organizations could play in promoting such schemes.
>
> There is no doubt that city or regional party committees can play a part in pushing an invention ahead despite departmental barriers. The drawback to this emphasis on the role of the local party organization is that some party officials are every bit as skilled at technological inertia as any ministry official. . . .
>
> Organizational innovations might nonetheless have some advantages. The gains from introducing them would, however, be limited as long as morale and incentives in industrial research and development are as low as they are currently being described in the Soviet press.[142]

The CPSU's efforts to stimulate scientific and technological progress have received very little attention in Soviet and Western writings.[143] Soviet commentators make general observations about the party's mounting importance, but its activities in specific spheres of modern science and technology are often unclear. Brezhnev apparently failed to increase the influence of innovative-minded party organizations over R and D conducted by state, production, and academic bodies. Cocks affirmed in 1982:

> Although the Brezhnev leadership has made some conceptual advances in reorienting policy toward technical progress, its efforts at restructuring the government to support the new policy orientation have run into problems of implementation. The leadership has found it extremely hard to recast the structure and attitudes of both a scientific and bureaucratic establishment that have taken decades to shape. More and more, the party's deepening involvement in scientific and technological development is directed at overcoming departmental problems and barriers and at prodding bureaucrats, scientists, and economic managers in the pursuit of technical progress.[144]

Because the USSR lacks a comprehensive science and technology policy and a cohesive set of institutions to implement this policy, one could conclude that it is premature or counterproductive for Soviet leaders to seek organizational and legal changes in R and D. But the adjustment of administrative rights and responsibilities is necessary, feasible, and pressing. Legally grounded integration of the present tripartite "system" was begun in the Brezhnev years in lieu of major changes in science policy and management. Also, interbranch integration was fostered by target programming, science-production associations, territorial-production complexes, and regional science centers. Dobrov, Rassokhin, Kostiuk, and others have tried to strengthen the legal bases of R and D, stressing that legislation can spur scientific and technological innovation but presently impedes it. R. O. Khalfina, a jurist, affirms:

> The legal regulation of management organizations and their activities creates the conditions for greater harmonization of departmental and general interests and for combatting attempts to place departmental interests above general ones. The need for strict observance of legality in all branches of state management, the establishment of effective sanctions when the law is violated, and the exposure of ineffective and illegal acts of management ensure the fulfillment of decisions and express the general interests of the state. The fullest use of these legal means will help to improve the level of management under conditions of the scientific and technological revolution.[145]

Burlatskii elaborates:

> Order and legality are prime requirements of the scientific and technological revolution, which presumes procedural stability in the making and implementing of decisions. The activity of the political system can be effective only given legality and order. This applies to *every* political institution. The importance of law and compulsion *grows* because of the need to overcome social cankers such as thievery, corruption, bribery, the violation of the legal rights of citizens, and other crimes.[146]

Burlatskii's analysis is grounded on the key constitutional provision (Article 6) stipulating that "all party organizations operate within the framework of the USSR Constitution." Also, Burlatskii implies that segments of the party, like those of other institutions, may be afflicted by or susceptible to the serious social ills noted. S. Kurits, in a more centrist and optimistic vein, concludes:

> Surmounting the contradictory and incomplete nature of the system of legal norms and improving management procedures should not be understood as another step toward the bureaucratization of work style. On the contrary, by singling out the officially required, routine part of work and, to a certain extent, formalizing the procedures for interaction among organizations and personnel subordinate to different departments—sometimes even assigning this work to a computer—procedural management, as an organic complement to economic methods of management, frees the time and mental energy of personnel to tackle nontrivial, creative tasks. . . .
>
> The need for managing on the basis of an improved system of legal norms and procedures for the preparation, adoption, and implementation of decisions has not yet been fully recognized. However, this is one of the most important means of shifting the economy onto the track of intensification.[147]

Hence, Soviet leaders and analysts have begun to examine the reciprocal influences of scientific-technological progress, management, and law and to understand that law is an important instrument of management in the era of the NTR.[148]

Expertise and Computers

Soviet theorists argue that "conscious" guidance of society can and must be enhanced in order to master the NTR and its multifaceted consequences.

They caution that not all consciousness is "scientific." Scientific management requires the active participation of technical specialists to help discover, apply, and accelerate objective laws and progressive tendencies.[149] As Afanas'ev argues, "The attempt to manage [society] on the basis of intuition will in time lead to subjectivism and is fraught with the danger of making insufficiently substantiated and ineffective decisions. Only by relying on diverse specialists can a leader conform to the demands of the time, the demands of the scientific-technological revolution."[150] As Fedoseev puts it, "A fundamental task facing scientific institutions and scientists is to . . . develop scientifically based recommendations that can be used in the practical work of central decision-making organs, ministries, departments, and public organizations."[151]

According to Soviet analysts, the social sciences are to play a major role in comprehending, controlling, and coping with objective processes. The characteristics and uses of knowledge depend heavily on the Marxist-Leninist orientation of the investigator. Even natural scientific knowledge cannot always be separated from the social conditions in which it is generated, and it can never be separated from the social conditions in which it is applied.[152] Hence, the principle of "partyness" (*partiinost'*)—an elemental faith in one-party rule and a sensitivity to the shifting priorities of the current CPSU leadership—is of considerable significance in scientific research. "A correct party approach" helps to orient the pursuit of objective knowledge, to ascertain its uses, to guide its practical applications, and to assess its benefits.[153]

Afanas'ev values highly the intuitive judgments of CPSU leaders *and* the scientific-technological and socioeconomic expertise of nonparty members. He stresses that the information needs of a complex and dynamic society place a premium on the ongoing reassessment and readjustment of decisions and ways to implement them.

> The management of a social system is a continuous process of the emergence and resolution of problem situations. Since there can be no management without social information, management—the resolution of problem situations—must involve the formation of a mobile information model in the managing subsystem. . . .
>
> The decision model must be mobile because the system is constantly encountering influences that seek to throw it off the course set by the program. If it is impossible or unfeasible to overcome these disorganizing influences by adjusting managerial actions, the program must be revised or a new decision taken. The range of adjustment or

> modification may be considerable, from small adjustments that barely affect the program to a full-scale reorganization of the program or adoption of a new one. . . .
>
> Adjustment is achieved by the manager's ability to "peep into the future," to compare current actions, current parameters with those prescribed in the program, to detect any deviation from the program indicating a need for correction. The social system's functioning is anticipatory; the manager must constantly have his eye on the future.[154]

Sophisticated expertise and technology can enhance the appearance or the reality of change. New approaches and technologies can create a pseudoflexibility and a pseudomodernity that actually block the periodic reexamination of goals and methods, impeding social and even technical rationality. The very same approaches and technologies, however, may be essential to socioeconomic progress.

Modern organizational technology does not necessarily rationalize the policy processes, to say nothing of the aims of political leaders. Powerful officials may not use quantitative information in their calculations and decisions. They may use only those quantitative data that support preexisting views and legitimize policy preferences based on narrow bureaucratic pursuits. The quality and practical value of quantified information may not prove to be particularly good, and top-level officials may be well aware of these shortcomings. Hence, even state-of-the-art technology and expertise may justify the unexamined intuitions of policy makers and help to perpetuate antiquated, ill-informed, and self-serving practices.[155]

Furthermore, Steven Marcus observes that an "expert system" consists of two basic elements:

> The "knowledge base" gleaned from the expert and the "inference engine," the logical structure within which a computer applies that knowledge in response to available evidence to draw conclusions and recommend actions.
>
> A major difficulty in building expert systems is that human experts are not usually aware of the exact mental processes by which they diagnose a problem; they apply their knowledge in subtle, seemingly instinctive, ways that elude codification. An effort is thus necessary to determine usable "heuristics," or rules of thumb, that can be embedded into precise logical sequences that a computer can execute.[156]

Such issues are quite relevant to the use of computers in Soviet policy making and administration. The Brezhnev leadership looked to computing as an important means of improving centralized planning and management, especially on economic questions and also in areas of social control. Nevertheless, a tension between ideologically prescribed science and empirical science remains a basic element of NUO theories. Soviet analysts try to reduce this tension by emphasizing both the subjective and objective aspects of management information. Also, they investigate "concrete" manifestations of "objective laws," rather than juxtapose empirical observations with unverified or unverifiable tenets. Science is to operate as the "eyes" of the societal guidance mechanism, while the political organs are to function as its "brain." Science is to be unfettered as long as it helps to implement the goals of the CPSU leadership. National goals are defined at least in part with reference to the technological possibilities immanent in scientific development. Scientists and engineers shape policy alternatives but do not make final policies.

Scientific management strives to enhance the capability of policy-making bodies to generate and utilize expertise. V. M. Grigorov declares: "In contemporary management systems the experience and knowledge of specialists must be as critical an element in the process of management as are technological means and organizational, economic, and mathematical methods. Only a harmonious combination of these components can create a highly effective system of management." The involvement of specialists in decision making helps to "scientize" societal guidance. Know-how, supplemented by quantitative information flows, is deemed important. Grigorov adds: "The solution of managerial problems is often achieved by using the judgments of specialists, whom leaders draw into the decision-making process. This is especially true when the performance of diverse managerial operations demands a great quantity of specialized but *unformalized knowledge and experience*. . . . The 'value' of collective experience in resolving different managerial tasks, particularly the 'value' of the experience of highly qualified specialists, is constantly growing."[157]

Policy makers are not to delegate authority to experts. Grigorov affirms, "A leader combines competence and power, as distinguished from specialists who possess only scientific competence. A leader's tasks do not automatically follow from the opinions of experts but are developed on the basis of a critical analysis of the conclusions and recommendations of experts."[158] The primacy of the politician over the expert derives from at least two sources other than their hierarchical positions. Effective decision making rests on an "optimal combination of specialized knowledge and experience in using this knowledge." The political leader translates the "abstract language" of science into the "concrete language" of decisions. Also, decision making is synthetic and goal-oriented, whereas expertise is compartmentalized and often value-free. Policy makers identify problems and opportunities, integrating knowledge from a wide variety of sources to resolve problems and to realize opportunities.[159]

Scientific management necessitates forecasting, which increases the significance of the expert. Dobrov and Smirnov maintain, "Without the direct participation of the most important scientists and engineers in forecasting work and without full representation of the viewpoints of various scientific schools of thought and their integrated forecasting projections, it will be difficult to produce trenchant and objective information about the future."[160] Forecasting combines imagination with specialized knowledge. Diverse organizational methods and incentives are needed to elicit forecasting data from experts and to transform such data into policy-relevant information.[161]

The Brezhnev administration, much more so than Khrushchev's, circumscribed the authority of forecasters. It placed forecasting knowledge in a subordinate and instrumental position vis-à-vis planning knowledge. Forecasting was said to be based on probabilistic, speculative, and perhaps irrelevant information, whereas planning was based on precise, concrete, and pertinent information. Although "forecasting is the cognition of processes," planning is "the form in which man's *influence* over these processes takes effect." Afanas'ev elaborates: "The forecast is not, as a rule, a directive in the form of a state decision. It is a recommendation or proposal, and it often comes in several variants. One variant is usually selected as the basis of a decision and assumes the character of a decision. It provides the foundation for the plan, which is a directive and contains specific targets and a timetable for their realization. The plan has the force of a state law and is supported by the appropriate resources."[162] Hence, forecasting and planning are linked, but planning is valued much more highly than forecasting. Political leaders who make planning decisions place themselves in a dominant position over experts who primarily contribute forecasting data concerning the "technological," rather than the "ideological," elements of socioeconomic plans.

The scientific management concept underscores the need for more and better communication channels between experts and policy makers. But scientific management does not legitimize criticism by experts of the values and goals that guide the utilization of expertise. Shakhnazarov, in emphatic and sweeping terms, declares:

> Neither in theory nor practice has scientific communism anything in common with a state system under which power belongs to the specialists as an independent political force. All official documents of the CPSU and the Marxist-Leninist parties of the other socialist countries invariably assert that until a full communist society has been built, society must be led by the working class and its revolutionary party.
>
> Clearly, there is no place here for a "dictatorship of specialists" or a technocratic regime.[163]

This argument is based on the assumption that technical rationalization does not necessarily lead to social rationalization. Indeed, technical rationality may result in policy legitimized by the inclusion of specialists in the policy-making process, regardless of their influence on the substance of the policies themselves. Experts may merely ratify the top leaders' decisions. But Brezhnev's and subsequent collective leaderships have often solicited expertise to clarify as well as to confront socioeconomic challenges, and specialists have thereby acquired circumscribed power to define those challenges and to pose alternative methods of meeting them. In a word, the relationship between the highest CPSU bodies and the scientific-technical intelligentsia is becoming more reciprocal.

Nonetheless, one must seriously consider the Stalinist legacy in assessing the impact of computers on Soviet decision making and implementation. Seymour Goodman observed in 1979:

> The introduction of computers into the structure of Soviet management has been limited. Conservative applications such as systems for scheduling, process monitoring, and personnel files seem to be the rule. Although there is some research on the utilization of computer techniques for decision analysis and for the modeling of management problems, little seems to be put into practice. Soviet managers tend to be older and more inhibited than their American counterparts. The system in which they work emphasizes straightforward production rather than innovation and marketing decisions. Soviet economic modeling and simulation activity stress the necessity of reaching a "correct socialist solution" and are not oriented toward being alert for general and unexpected possibilities in a problem situation. Furthermore, Soviet industry has learned not to trust its own statistics; much information is controlled and unavailable to many potential users, and there may be a big difference between "official" and actual business practice. What does one do with a computer system for the "official"

> operational management of an enterprise when actual practice is different? Does one dare use the computer to help manage "expediter" slush funds, under-the-counter deals with other firms, etc., which could leave a substantial trail for audits?[164]

Aron Katsenelinboigen, a mathematical economist who emigrated from the USSR, stated in 1980:

> During the last twenty years, experience has been gained in introducing economic-mathematical methods at various levels of the Soviet managerial hierarchy. Presently, under pressure from the country's leaders, the automated management systems (ASU) are being introduced into factories. These systems, based on primitive mathematical models, are part of the present planning technology. The immediate effect of the ASU is not great, and sometimes it is even negative because of the increased expenses for servicing computers. Apparently, the effect of computers was basically indirect. . . .
>
> The introduction of automated management systems at the higher levels of management comes up against great difficulties. Various trends in the development of Soviet economics cannot but touch upon the interests of managers on different hierarchical levels. Thus, economic-mathematical methods require a great deal of additional knowledge, which the manager often cannot master. The introduction of these methods also might lead to a partial reduction of the managerial personnel, a problem that authorities face in overcoming resistance.[165]

In short, Soviet bureaucrats and production executives have good reason to be chary of technological innovations. Incumbent party officials at all levels are especially cautious in utilizing modern information processing technology because of the high probability of disrupting the status quo. The introduction of computers is most likely where bureaucratic units are dissatisfied with existing conditions and are urgently seeking technological or ideological means of advancing their interests. Under Brezhnev, however, officials enjoyed unprecedented job security. Under Andropov and, to a lesser extent, Chernenko, the campaigns against corruption, waste, and sloth were hardly conducive to the computerization of information flows.

Furthermore, the fragmentation of the Soviet bureaucracies has impeded the design, manufacture, installation, and dissemination of computer technology, software even more so than hardware. N. C. Davis and Goodman affirm:

> Like most other sectors of the Soviet economy, software production has to contend with a major behavioral obstacle. Soviet organizations with similar interests tend not to cooperate or interact with each other. Tradition, institutional structure, and incentives are such that enterprises try to tend to their own affairs as much as possible. Much of the cooperation that does exist is forced by party or military demands or by desperate efforts to circumvent supply mistakes. Other efforts at cooperation are rarely effective. This has particularly affected software diffusion. Before the existence of Riad, hardware manufacturers did little to produce, upgrade, or distribute software. Few models existed in sufficient numbers to make a common software base of real economic importance possible. Repeated attempts to form user groups came to little. Soviet security constraints restricted those who could share software for some models, and enterprises rarely exchanged programs.
>
> Thus, the population of experienced programmers remained small in the USSR. This was compounded by the failure of the Soviet educational system and computer manufacturers to provide the kind of hands-on, intensive practical training that is taken for granted in the U.S.[166]

The Soviet demand for computer analysts has increased considerably and their training has improved since the early 1970s. But computer specialists are not likely to develop a shared group "interest" or the capacity and will to promote it any more than, say, economists and industrial managers. Rather, individual officials and bureaucratic units are using quantitative data and data processing expertise for all kinds of particularistic (and possibly a few nonpartisan) purposes. The advent of the computer age, especially the dramatic development of microcomputers, personal computers, and word processors, could easily spawn or enhance new forms of departmentalism and localism.

Glushkov was probably the most forceful Soviet advocate *and* the bluntest critic of computerized communication in the USSR. He contended in 1982 that the "chief content" of the NTR was "the appearance of an essentially new man-machine technology for processing information. Since the circular flow of information is the basis for the functioning of any organization, the NTR must be viewed primarily as a revolution in organization and management."[167] At the same time, Glushkov was keenly aware that computers were merely components of larger political-administrative and socioeconomic systems and that computers could help to organize, direct, and control only with substantial changes in the communication patterns of these larger systems. According to Glushkov, it was *impossible* to improve economic management and to spur scientific-technological progress with the traditional Soviet information processing equipment and practices; without structural and behavioral changes in the economy, the rates of economic growth and productivity would increasingly *decline* in the era of the NTR.

> The conservatism of the traditional technology for processing planning and management information leads to the intensification of "disorganized complexity" in the national economy and erects informational-organizational barriers to planned economic growth. The deterioration of the quality of management, the deadening of a larger and larger portion of planning information, and the amassing of "dead time" in the movement of primary information and feedback are beginning. Consequently, there is a loss of growth and reduced rates of economic development.
>
> The problem, of course, is not just in the technology of organizational management. The economic mechanism plays a large (indeed, a primary) role here. . . . However, it is important to emphasize that economic mechanisms (especially under socialist conditions) "do not work" by themselves in isolation from the organizational-management system.[168]

Glushkov bitingly remarked in 1975 that automated control systems were chiefly hampered by the "half-truths" entered into computers, and that "we find ourselves somewhere between confusion and a search for scapegoats."[169] Also, Glushkov was keenly sensitive to the impact of the Soviet economy on the production and deployment of computer hardware and software. As Allan Kroncher contends, "One major problem is the planning requirement that computers be produced in specified quantities regardless of whether they meet the latest standards of technology and with no allowance being made for the qualitative advantages of one machine over another." Moreover, Kroncher notes the importance of bureaucratic considerations. According to *Pravda*, one automatic control system was dismantled and sold because it "impartially pointed out management's blunders and omissions."[170]

Soviet computers are parts of larger social and organizational communication systems, most of whose members resist innovative modes of analysis and new technologies. "Changes in high-level perceptions and technological development . . . are no longer the critical, limiting factors they once were," Goodman argues.

> The real problem is now more broadly and deeply systemic: the pervasive and effective integration of computing into the fiber of the national economy. . . . Development and use of computers on a national scale cannot be isolated socially like the more narrow military industries, or even like industries such as power, steel, etc., whose products are used relatively passively as compared with computers. In particular, it might not be possible to contain major reforms that would greatly improve the quality and availability of computing services. If changes were made to enable a Western-style service sector to exist among thousands of computer installations cutting across the domain of every ministry, then (if successful) how could there not be pressure to extend these changes beyond computing?[171]

Succinctly stated, computing is least likely to change the Soviet leaders' cognitive orientation and basic values, more likely to change policy-making procedures and attitudes, and most likely to change substantive policies and beliefs. Although better information makes possible better-informed policies, it creates more choices and accountability and threatens firmly established formal and informal bases of power throughout the entrenched bureaucracies. The NTR has not had a considerable impact on Soviet decision making and implementation, but especially because of the burgeoning worldwide use of small and mobile computers linked and unlinked to ever more powerful mainframe computers, important changes are possible in the not-too-distant future.[172] The *nature* and *extent* of these changes and their effects on Soviet management will depend on the interaction of many political-administrative, socioeconomic, and scientific-technological factors. Brezhnev's successors are striving to control the adaptation of traditional organizational practices to new technologies, and their capacity to do so, while coping with mounting economic and social problems, is of enormous significance for the future of the Soviet polity.

Unresolved and Contentious Issues

One senses that the official Soviet optimism about the NTR stems in large part from the belief that it is an important source of power to be harnessed. But the actual uses of this power, the ways in which choices are to be made, and the evolving structure of political influence in the USSR are among the fundamental issues that are rarely discussed by Soviet theorists. As we have seen, the relationship between the party-state and the people is handled in a particularly circumspect manner. Soviet commentators focus heavily on "goal-*seeking*" rather than on "goal-*setting*" feedback—even the Afanas'evs and Burlatskiis, who recognize the benefits as well as the risks of both kinds of information.

Soviet analysts have predicted that the NTR will produce major changes in scientific, economic, and managerial thinking. Afanas'ev asserts: "The computer today is not only a powerful calculating instrument and means of processing and storing vast quantities of information, but it is also a means of human intellectual activity. Under the influence of computers *new structures of human thought are formed*, the organization of physical and intellectual labor, cognition and learning, the spiritual world of man, are changed."[173] Nonetheless, Soviet observers rarely explore how changing cognitive processes may influence political calculations or how changing social and bureaucratic "interests" may affect policies and policy making in different issues areas.

Although a common theme is that the NTR is bringing improvements in economic planning and management, the CPSU leadership's primary purpose has been to strengthen the present centralized economic system and to achieve traditional goals in the most "rational," "effective," or "optimal" manner. Such pragmatic concepts, however, encourage discussion of *various* ways to improve the economy, *various* standards and criteria to judge the effectiveness of these recommendations, and *various* conceptualizations of "scientific management," "centralism," and "democracy." Hence, the instrumental focus of Soviet theorizing invites debate about political-administrative aims, priorities, and ends-means relationships. The actual and probable impact of most policy-making procedures, analytical approaches, and technologies are not self-evident and, like organizational goals themselves, are frequently in dispute.

Notwithstanding a common Marxist-Leninist world outlook, leading Soviet officials and theorists have expressed competing ideas about "scientific management" since the last years of Stalin's rule. On the one end of the spectrum is the traditional industrialization strategy, which includes autarky, autocratic centralism, command planning, and military and heavy industrial priorities. On the other end of the spectrum is the advanced modernization strategy, which includes selective interdependencies with the global community, authoritarian centralism with specialized elites participating actively in decision making and implementation, optimal planning systems responding to market forces, and balanced military and overall economic development utilizing science-based high technology in defense and civilian industries. All along this spectrum are shifting combinations and compromises, intellectually and politically inspired.[174]

For example, there was an ongoing debate in the Brezhnev years about the role of the CPSU in the economy, and there were distinct "engineering" (maximization) and "managerial" (optimization) perspectives about economic planning, decision making, and organization, growth and productivity, and professional training of newly recruited and coopted party cadres. These two perspectives were epitomized by the centrist and reformist orientations of Brezhnev and Kosygin respectively. Brezhnev stressed the need to improve the efficiency and quality of production in existing industrial and agricultural facilities; Kosygin called for improved production at existing facilities *and* for new civilian sector factories and high quality, exportable goods. Brezhnev implied that economic growth and productivity are primarily "political" problems and necessitate considerable intervention in economic matters by *party* organs at all levels; Kosygin emphasized that growth and productivity are complex political, economic, and social problems involving the legal obligations of *state* organs to fulfill plans and to "manage" the economy. Also, Brezhnev was more sanguine than Kosygin about the extractability and processing of Soviet natural resources, a difference that probably contributed to the two leaders' contrasting ideas about East-West economic relations and about military versus civilian investment priorities.[175]

Different Soviet assessments of economic performance and its political implications stimulated different responses to proposed changes in the structure and functioning of the economy. Modern approaches to planning and management, such as systems analysis, were embraced by many members of the scientific and technical intelligentsia but only by some party and state officials, especially at the national level. Without modified professional and personal incentives, why was it in the best interests of, say, a republic CPSU first secretary to promote technological innovations in factories or farms? Even if one were so inclined, could one induce local factory managers and collective farm chairpersons to follow suit? Hence, the effects of the new methods associated with scientific management and with alternative conceptualizations of a mature socialist society depend on much larger political choices and on bureaucratic and interpersonal conflict and cooperation.

Moreover, by what criteria and standards are systems approaches and innovative organizational technologies to be judged? Is immediate, medium-term, or long-range assistance to be given to "rational" decision making in certain organizations and sectors? Which units and priorities are to be used and for what kinds of output and outcomes? How and by whom is productivity to be measured and rewarded? The answers to such questions were, are, and will remain highly political. This is especially true in today's increasingly interdependent, complex, and dynamic internal and external environments, where credit and blame can be justifiably or arbitrarily assigned to various organizational units, domestic and foreign.

Soviet theoretical and empirical studies of the major bureaucracies expanded during the Brezhnev years and thereafter. Leading CPSU officials have cautiously encouraged innovative theorizing about "the political organization of society." They have also called for more interdisciplinary team research on socioeconomic and organizational dilemmas and fuller investigation of informal practices because "various records and reports do not always reflect actual processes correctly."[176] Furthermore, party leaders have criticized social scientists for avoiding subjects of *pragmatic* as well as theoretical significance, such as ineffective decision making and implementation, departmentalism and localism, bureaucratic malfeasance and unaccountability, and indifference to Marxist-Leninist ideology. Although many empirically based Soviet studies of administrative behavior and public opinion are probably classified by the party and state agencies that commissioned them, a growing number of articles and books focus on the major bureaucracies, not merely on the frequently studied mass organizations. Hence, Soviet research is generating more and more empirical data about specific institutions, processes, and systems and is becoming less and less constrained by either an a priori ideological approach to unresolved problems or by a primarily formal and legal-constitutional mode of analysis.

CPSU spokespersons usually avoid or obfuscate the issues of power, institutional reforms, and goal-changing feedback, but they are increasingly discussing the *actual* and *distinctive* features of Soviet political-administrative behavior and are trying to develop an amalgam of the most useful socialist and nonsocialist organizational technologies. Indigenous and imported know-how and machinery are to be assimilated into the existing institutions and changing social order. Public communications are to play a significant part in disseminating the enormous quantities of normative and technical information needed to guide a more and more industrialized and differentiated mature socialist society. The interplay between the Soviet polity and the mass media is especially important and is the theme of our next chapter.

4
The Mass Media, Communication Systems, and Dissent

The Soviet mass media were designed to unify the population during the industrialization of a relatively underdeveloped society. Under mature socialism, however, unity is not generated or sustained merely by molding new industrial laborers. In addition, the different interests and opinions of an increasingly well-trained and specialized labor force are to be articulated and aggregated. The present-day CPSU leadership uses the media to forge unity and to provide a forum for diverse viewpoints on selected issues. These dual purposes conflict with one another in many ways.

The significance of Soviet mass communications is underscored by the fact that the editors-in-chief of the three major newspapers—*Pravda*, *Izvestiia*, and *Sovetskaia Rossiia*—together with the head of the state broadcasting committee, regularly participate in the weekly meetings of the CPSU Secretariat, the most powerful political body in the USSR after the Politburo.[1] Afanas'ev, editor-in-chief of *Pravda*, affirms: "The mass information media play a large role in introducing and solving the most important economic, sociopolitical, scientific-technological, and ideological problems. . . . [The media under contemporary conditions] do not merely record events and discover the state of affairs, but actually intervene in events and guide their course in conformity with the public interest." Developing this theme, Afanas'ev presents an authoritative summary of the aims and operations of the Soviet media:

> The principal functions of the mass information media are *political-educational*, *propagandistic*, *managerial*, and *organizational*. They make people aware of the party's and state's policies and of management decisions, orient people to the countless number of social events, report and comment, enlighten and convince, recommend and appeal, motivate and organize the masses, shape public opinion, and strengthen

> a system of social values and patterns of behavior. On the other hand, the mass information media are a means of ascertaining the attitude of the masses toward party and government decisions and events at home and abroad and are channels for communicating proposals and critical comments, as well as signals concerning the functioning of different levels of the social organism and of different management organs and leaders.[2]

The complex and diverse roles of the Soviet media are enhanced by the broadening and deepening of participation in decision making and implementation, especially by bureaucratic, professional, and social groupings. "Proposals and critical comments" are likely to increase in direct proportion to the quantity of management decisions, to their stages and levels, to the range of "problem situations" acknowledged by the leadership, to the interconnections among policy areas, and to the number of administrative bodies carrying out inconsistent or conflicting directives from the center. Soviet bureaucratic, scientific-technical, and cultural elites use the press to mobilize support for desired policies and for preferred means of policy implementation. Attempting to meet their legally binding obligations and to defend their professional and personal prerogatives, party and state officials debate the efficiency, effectiveness, and even the wisdom of national programs through public as well as private communication channels.

These themes and others will now be discussed with emphasis on the Soviet leadership's responses to socioeconomic trends and telecommunication breakthroughs at home and abroad.

The Communication System of State Socialism

Soviet perspectives on the media developed in the context of the challenge to overcome historical backwardness rapidly. The USSR's communication system was constructed for the purpose of facilitating political, social, and economic progress. New communication strategies were to help modernize the political culture and institutions, the social structure, and the economy. A backward society was to be propelled into the modern age, and the party was to guide the trajectory with a forceful boost from mass communications. Above all, illiteracy was to be eradicated, and socialist values were to be inculcated.

Rapid modernization in Soviet Russia was intensely politicized, and the media were instruments of class politics. Because CPSU leaders deemed the proletariat a revolutionary class and claimed to understand the proletariat's "genuine" interests, they viewed the media as tools to help create a cohesive social consciousness. Also, the traditional Soviet approach to the media was predicated on the international isolation of the USSR. Stalin's famous tenet, "socialism in one country," was rooted in the self-sufficiency of Soviet society. The communication system was to spur industrialization and collectivization rather than to expose citizens to alien influences from "class enemies" in the USSR and in the West.

According to Stalin, fast-paced economic and military modernization was essential because the outside world was hostile to the Soviet Union. This message was incessantly communicated to the masses, and its essence is contained in one of Stalin's best-known pronouncements: "To slacken the pace would mean to lag behind; and those who lag behind are beaten. We do not want to be beaten. . . . [Old Russia] was ceaselessly beaten for her backwardness. . . . We are fifty or a hundred years behind the advanced countries. We must make good this lag in ten years. Either we do it or they crush us."[3] In such an approach to modernization, there is no room for alternative viewpoints. During the purge trials of the mid-1930s, prosecutor Andrei Vyshinskii denounced the former right opposition as "enemies of the people." Even the slightest variation from the regime's definition of the truth was considered a manifestation of backwardness, disloyalty, or treason. The media were to communicate homogeneous directives from the leaders to the led and to help create "monolithic" political and social unity in the face of internal and external threats, real and imagined. Socioeconomic accomplishments and national security were thought to be the products of enforced consensus, which the media were to strengthen and sustain. Hence, ideological campaigns were a vital weapon in the struggle to establish and develop a socialist order.

In the state socialist polity, authority rested on three philosophical planks. First, Soviet theorists maintained that "positive freedom" is superior to "negative freedom." Under liberal democracy, liberty is negative in character and consists of freedom from interference by other persons, groups, and institutions. Positive liberty, however, consists of freedom to transform the world in conformity with a desired conception, identity, or purpose. Political leaders posit a common good and potential self against which existing conditions are found wanting. Positive freedom is based on the assumption that people should be coerced to pursue goals that "they would if they were more enlightened themselves pursue, but do not, because they are blind, ignorant, or corrupt."[4] Second, citizens under state socialism were to develop the perspectives and skills that advance the national interest. The exercise of freedom was viewed as the *collective* ability to transform existing selves into higher selves. The new socialist men and women could fulfill their potential only as parts of the whole society and by working unselfishly for the welfare of all. Third, individual pursuits were equated with selfishness. Accordingly, neither the individual nor society was thought to benefit when the individual criticizes the political system from his or her vantage point. Because history, state socialism, and the CPSU were presumed to be developing "progressively," rationality was a collective rather than an individual attribute.

Political power was legitimized in state socialist systems on the basis of these principles. Claiming to be the "vanguard of the proletariat," national leaders monopolized power in order to subdue particularistic interests that could triumph over the public interest. Ideologists argued that the society and polity must be integrally connected and that they must be guided and coordinated by the leading political organs. Conflict in the making and administering of policy was to be strictly regulated by Stalin personally or by the party leadership. Feedback and primary data were necessary only to the extent that they strengthened centralized control and facilitated the implementation of the goals and policies proclaimed by the executive institutions. The mass media's role was to assist CPSU leaders, all of whom were purportedly "conscious" of the laws of social development and of the continuing, sometimes heightening, "spontaneity" of people and impersonal forces. In a word, mass communications were key mechanisms through which central institutions exercised positive freedom vis-à-vis society.

Under state socialism, the general will was a primary value. Whereas the fragmentation of interests in liberal democracy was viewed as a manifestation of historical decay, the unity of interests in socialist democracy was considered essential to progress. Also, centralized decision making was to fuse a monopoly over knowledge with a monopoly over wisdom. Information was to gravitate to the center because top CPSU officials insisted that only they were capable of using this information to formulate and implement just and effective policies. Furthermore, the state ownership of the means of production was to ensure the dominance of the general interest over selfish interests. Socialization was equated with public control over individual actions and the supremacy of the selfless over the selfish and the rational over the irrational.

The idea that the media are tools to be utilized by the party and state is probably the central attribute of the traditional Soviet view of the media's role.[5] By helping to construct and develop a socialist society and by laying the foundations for the future communist order, "The socialist press, films, radio, and television are powerful *instruments* in the ideological and political education of the workers. They help to form the workers' *communist attitudes toward work, communist ideology, and morals*."[6] Also, Soviet commentators affirm that journalists must support the party line, adhere to Marxist-Leninist ideology, convey information truthfully, generalize and disseminate political information, serve the public, and constructively criticize shortcomings in the implementation of national policies.[7] In short, the communication model of state socialism strongly emphasizes the propagation of a cohesive ideology and the establishment of a homogeneous industrial society. But, as industrialization progresses, serious challenges to this model emerge.

The Communication System of Developed Socialism

The communication system Stalin established in the 1920s and 1930s has been placed under increasing strain during the transition from socialism (1936) to developed socialism (1967) and thereafter. To analyze the sources, nature, and consequences of the strain on Soviet communications, we find it useful to view the USSR's maturation in the following way.

Contemporary Soviet society is exerting more and more pressure on the party-state. As the polity becomes more differentiated, flexible, and permeable to internal and external influences, it creates more axis points for organizational units and social groups to express their interests and opinions. The rationalization of management and the rebirth of civil society are twin pressures associated with mature socialism. Both of these pressures mounted in the Brezhnev period through the expanded participation of experts in the policy process, the CPSU leadership's efforts to meet rising consumer demands, and the initiatives from above and below for improved two-way communication between the bureaucracies and the public. The traditional communication system, which was chiefly an instrument for maintaining political control and for transmitting commands from the center to the periphery, became a stimulus for the revitalization of mass participation in decision making and implementation and a channel for transmitting information from the citizenry to the leadership. Indeed, Soviet analysts disagree about whether the mass media are primarily a segment of the polity or of society.[8]

The generation and transmission of knowledge under rapidly changing socioeconomic conditions place a dual tension on the Soviet communication system. On the one hand, the party guides society by means of mass communications, a traditional function of the state socialist model. On the other hand, the successful guidance of an industrialized society requires greater inputs from experts, who are to provide CPSU leaders with data and feedback about the feasibility as well as the consequences of policies. The mature socialist communication system is to reflect and disseminate such inputs, a process that can undermine the party's directive role.

Illustrative of this problem is the treatment of conflict in the concept of developed socialism. On the one hand, greater cohesion in advanced society is to be nurtured. People and nature are to be unified on a higher scientific-technological and socioeconomic level.[9] On the other hand, the diversity of goals and needs produced by a highly industrialized economy, the expanding participation of specialists in decision making, and the growing significance of feedback in the shaping and reshaping of policies —all of these trends underscore the heightening conflict of interests in a mature socialist society. The absence or paucity of "antagonistic contradictions" notwithstanding, there is ample conflict among "nonantagonistic contradictions." The overcoming of nonantagonistic contradictions, such as the tension between the rising demand for consumer goods and the continuing heavy industrial-defense priorities, is crucial to progress in developed socialism.

Soviet mass communications were designed and constructed to foster unity, not to help resolve conflicting interests and opinions. The mounting structural differentiation of Soviet society considerably affects journalists, for example, who are charged with the responsibility of discerning and defusing the nonantagonistic contradictions exacerbated by the NTR. Journalists are to assist party leaders in "overcoming a whole series of contradictions. . . . The more sharply the journalist focuses on and describes conflict situations, the greater his responsibility for recommending methods to solve these conflicts."[10]

Although the Soviet media strive to promote a consensual ideology, they must at the same time help to investigate and ameliorate problems and to manage the competing aspirations and viewpoints of bureaucratic and social groups. According to an authoritative Soviet text, "The differences of opinion in socialist society do not touch the fundamental issues of ideology; they are concerned with certain, as yet unsolved, questions of social life that require discussion."[11] Unsolved questions, however, are burgeoning in present-day Soviet society. The Brezhnev administration increased the opportunities for debate within the Soviet press in order to vent elite and mass discontent on selected issues. But the CPSU leaders' efforts to elicit more and better feedback and primary data were portentous. Many of these efforts intensified pressures for policy changes, and a few may have been disruptive. Ellen Mickiewicz observed in 1981 that the Soviet leadership "has claimed to know the attitudes of its population without the need to survey it. These claims have been greatly weakened in the past decade or so with the advent of public opinion polling."[12] Although much survey research is classified, some has been conducted by or reported in the media, especially the press.

Briefly stated, the traditional communication model exists side by side with the modern one, and each influences the other. The traditional structures have been modified to perform both the old and the new functions. For example, journalists must somehow strike a balance between their responsibilities for the ideological molding of the population and for the unearthing, channeling, and interpreting of policy-relevant information. The former role is still predominant, but the latter is of increasing significance.

Several developments have undermined the monopolization of the Soviet communication system by a homogeneous political elite, which tries to steer society in an ideologically determined direction. Occupational specialization and bureaucratization have reduced the capability of the national leadership to formulate feasible policies and, above all, to implement them effectively and efficiently. Also, the audiences receiving media messages have become more socially and economically heterogeneous, and, because of public opinion research and investigative reporting, CPSU leaders are acquiring a much better understanding of this heterogeneity. Furthermore, the isolation of Soviet society from the external world has declined, primarily because of advances in telecommunications and changes in the world economy. Finally, global scientific-technological and socioeconomic changes have altered communication patterns in the USSR. Let us examine these four developments in turn.

Political-administrative Dimensions. Since Stalin the differentiation of elites and the fragmentation of bureaucratic units have spawned specialized media that articulate the interests of policy groups. In turn, group awareness has been enhanced by the increasingly specialized nature of the Soviet media, through which actual and incipient coalitions communicate to one another their views on political and administrative issues. Soviet officials exchange ideas and information in journals and newspapers and (to a much lesser extent) on radio and television. Most important, they do so for specific purposes, such as mobilizing bureaucratic and specialist elites to support or resist policy changes under consideration in top CPSU bodies or to implement, adjust, or discard a faltering program.

To be sure, policy making is highly secretive in the USSR, and Lenin and subsequent Soviet leaders have maintained a facade of "monolithic unity," in accordance with the party rule forbidding discussion of policy alternatives in CPSU meetings and in the press *after* a party organization has reached a decision. Brezhnev and his colleagues preferred private rather than public dissemination of policy-relevant information, and they relied heavily on the central and regional party committees to study and evaluate policy alternatives and personnel, as well as to coordinate and supervise policy implementation. But the collective leadership's communication policies increased information flows within and among the party, state, and mass organizations and the citizenry. Brezhnev's Politburo consistently encouraged greater publicity, openness, and frankness (*glasnost'*) within parameters set and changed at will by top CPSU leaders.

Stalin's successors have continuously tried to influence one another's thinking and behavior through public as well as private communications. Diverse policy groups have used the media to initiate and resist incremental changes in a wide range of issue areas. As Timothy Colton affirms, the Brezhnev leadership accelerated Khrushchev's initiatives toward "fuller public airing of policy questions."

> After 1964 the political elbowroom of the skilled professional was greatly expanded in nearly every field of policy. Politicians could still veto change, declare ultrasensitive matters out of bounds, and control the way ideas were expressed. Yet, the richer and blunter discussion of policy issues, big and small, should impress anyone reading Soviet publications of the day. Advisors in many fields were permitted and urged to do public opinion surveys and incorporate the results (which were not always published) into their recommendations. Ironically, the same regime that locked dissidents in psychiatric hospitals and outlawed proposals for radical change also countenanced the most candid debate seen in Soviet Russia since the 1920s on a host of within-system issues: in the economic sphere, on questions of planning, investment, manpower, energy, trade and technology transfer, and technical innovation; on institutional questions, the relative power of central and local agencies, the handling of citizens' complaints, and the role of the trade unions; in the social area, problems of crime, pollution, the needs of the elderly and of working women, the difficulties of declining villages and of metropolitan areas; or simply on aspects of life style such as family size, alcoholism, and the effects of television.[13]

The importance of these observations cannot be overestimated. Policy-relevant debate in the mass media *increased* under the Brezhnev administration. The reluctance of Westerners to acknowledge this fact has diminished our understanding of the USSR's development and has impeded our ability to study it.

Socioeconomic Dimensions. The changing composition and needs of Soviet society have also affected the media. In 1964 the CPSU abolished limits on subscriptions to newspapers and journals. Not only did circulation increase, but the choices of readers became an important consumer pressure affecting the communication system. Soviet newspapers initiated surveys of the characteristics and concerns of their subscribers in order to ascertain and respond to readers' preferences.

These surveys revealed a heterogeneous audience. Human interest stories, international news, and articles presenting different viewpoints on the role of factory managers, bureaucratic red tape, and local socioeconomic conditions were among the most popular items. Theoretical and propagandistic articles were among the least popular. The more highly educated segments of the population were more likely to read newspapers and journals than the least educated segments. Political and economic materials tended to be read primarily by persons who have a higher occupational status. Nonparty readers rarely turned to party-oriented articles first, preferring articles on international politics, general themes, sports, science, culture, and economics. Not surprisingly, CPSU members read the articles on party affairs before turning to the subjects listed above. Surprising and distressing to party leaders, however, "it was found that education was inversely related to agreement and satisfaction with the media."[14]

Surveys of this kind were conducted by recently trained Soviet specialists in public opinion research, who were often employed by party and state agencies on a full-time or consulting basis. Many of these sociologists encouraged the media to eschew distorted images of the public mood and to help identify and resolve dilemmas. Discussing the activities of Soviet survey researchers, Mickiewicz comments: "Although they do not fail, in their writings, to support the guiding role of the [CPSU's] communications experts, their research is directed primarily to portraying public opinion, demonstrating the incongruence between that picture and official assumptions and policies, and, finally, recommending ways in which the political elites must alter their behavior."[15]

The increasing differentiation of Soviet society has been accompanied by the increasing differentiation of the media. Interpreting the findings of a poll of *Izvestiia* readers, two Soviet analysts concluded: "Radio presents the news almost immediately; the newspaper, on the average for the country, within three days; and the magazine brings it no sooner than a fortnight. Yet editors sometimes assume that they alone are supplying the reader. . . . Today, more than ever, *each publication is required to have its own individuality, its own characteristic selection of information. . . .*"[16] Also, a heterogeneous audience constitutes a "market pressure" on the communication system, inducing the media to meet consumer needs more adequately. Each newspaper is to focus on a specific audience, taking into consideration the readers' profession, educational level, and political maturity.[17] Furthermore, Soviet theorists consider media specialization to be a manifestation of democracy in a developed socialist society. According to M. V. Shkondin, "The creation of a press that is oriented toward the diverse groups and social strata in the USSR is a basic achievement of socialism and the realization of the Leninist idea of the freedom of the press."[18]

International Dimensions. The insularity of the Soviet communication system is being broken down by several external developments.

First, the impact of global communications on Soviet society makes isolation difficult to maintain. Zhores Medvedev, a dissident scientist, argues:

> In the course of the last two decades science and technology have created massive means of communication and contact between peoples, which have united the whole world into a single information system. Worldwide radio and television, the development of national broadcasting to other countries, the propagation of newspapers and journals, and the rapid growth of international tourism and trade have created a mutual understanding and mutual influence of nations and countries as an observable fact. The citizen of almost any country who is sufficiently intelligent can now form his view of the world from the information which he receives from various sources. If there is not enough information from his own country, he can borrow from other sources. This leads to a society which is ideologically and politically heterogeneous and makes the governments of countries tolerant of different points of view in the population. This kind of tolerance does not always appear without friction, and in the events in our own country, for example, we can see attempts to preserve an abnormally monolithic ideology and opinions. . . . But the general trend is under way in our country, too.[19]

Second, the increasing participation of experts in policy making and administration enhances the importance of the Soviet intelligentsia, whose work can benefit from greater access to foreign sources of information. Soviet intellectuals are eager consumers of Western news and data. L. G. Churchward notes:

> Despite close control over international contacts, Soviet intellectuals are often surprisingly well informed about international developments in their own and allied fields. Students and graduates alike listen regularly to foreign news broadcasts, including the British Broadcasting Company, the Voice of America, and Radio Liberty. To meet the interest of Soviet intellectuals in overseas news, the Union of Soviet Journalists was forced to establish a special weekly paper, *Za rubezhom*, consisting mainly of translations from foreign newspapers. This paper has a circulation of almost a million and is always quickly sold out.[20]

Third, the Soviet population's appetite for Western communications, especially the scientific-technical intelligentsia's demand for both printed material and radio broadcasting, places a market pressure on the Soviet system. Journals such as *Literaturnaia gazeta* and *Zhurnalist* have Westernized their writing and layout styles in order to make them more aesthetically pleasing, informative, and persuasive. Also, the Soviet media have traditionally been very slow in bringing news to the Soviet people, especially information displeasing or inconvenient to the party-state leaders. But many of the Soviet nationality groups have access to such information through Western broadcasts in their native languages. Although Radio Liberty was consistently jammed throughout the Brezhnev years, the Voice of America, the British Broadcasting System, and Deutsche Welle were not jammed between June 1963 and August 1968 and between September 1973 and August 1980. Even with jamming to major cities, some foreign broadcasts reached Soviet suburbs, cities under 500,000 people, towns, and rural areas. Radio Liberty claims a stable listenership of 7 percent of the adult Soviet population, especially among the 30–49 age group, men, citizens with higher and secondary education, and urban residents and especially in the Ukraine, Belorussia, the Caucasus, and the Baltic republics.[21]

Fourth, Western broadcasts have shaped the perspectives of some Soviet officials. According to a Soviet newspaper editor, "It is very important that people are informed in good time and correctly about all events taking place both in our country and abroad. There must be no gaps in the information because these may be used by ideological enemies."[22] By Western standards, gaps and time lags in Soviet news reporting, especially the reporting of unpleasant news, remain common. But the Brezhnev administration authorized editors to speed the publication of current events, subject to approval by censors in the editorial offices, and the contrast to previous Soviet practices was considerable.

Advanced Industrial Dimensions. Several characteristics of the Soviet communication system are directly affected by the emergence of an advanced industrial society in the USSR.

First, there has been a communications explosion in the post-Stalin period. Between 1950 and 1973 the yearly circulation of newspapers increased five times, journals seventeen times, and books almost twice, with national and regional newspaper circulation exceeding 175 million in the early 1980s. In 1960 there were 4.8 million television sets in the USSR, but in 1976 there were more than 55 million, with more than 7 million manufactured annually.[23] Soviet mainframe computers numbered about 120 in 1960 and rose to about 22,000 in 1978, with U.S. figures jumping from about 5,500 to about 200,000 in the same period. The quantity and quality of Soviet telephone communications remain a serious problem. Noting that telephones "have become an essential precondition for the widescale use of many modern means of information technology," Julian Cooper reports that "the number of telephones installed per 100 people in the USSR in 1982 was at the level achieved in the United States by the late 1930s."[24]

Second, two-way communication between the party-state and the public has become increasingly important. As R. A. Safarov emphasizes, "The mass information media must shape and reflect, inform and be formed by, public opinion. These tasks are dialectically interconnected. . . . *Thus, it is impossible to shape public opinion effectively without being informed by it.*"[25] Or, as Chernenko declared in 1983:

> It is impossible to conduct effective ideological work without a precise feedback mechanism. Otherwise propaganda will only spin on its wheels.
>
> The party has many channels of flexible, effective communication at its disposal that allow it quickly to pick up changes in popular sentiments and to study in depth the masses' interests and needs. These include the comments and letters of working people, questions to lecturers and propagandists, and sociological research. . . . [But] we must switch from appraising the state of ideological processes to actually forecasting them, from random studies of public opinion to systematic polls, and *perhaps we must even set about organizing a center for the study of public opinion.*[26]

Third, strategically significant information remains a scarce resource that is closely linked to status and power in the USSR. Control over key information is an important privilege dispensed by the political elite, and the information people receive and the officials they communicate with depend largely on their professional positions. But Soviet society is experiencing a "revolution of rising expectations"—a mounting demand for up-to-date information rather than "timeless" stereotypes. Soviet survey research demonstrates that media audiences are interested in the news rather than in ideological truisms. Fresh information whets people's appetites and motivates them to study political issues and current events.

Fourth, the technical and normative information needed to perform professional roles is increasing, and the media are to communicate such data and expectations throughout a more and more complex society. For example, the Brezhnev administration attempted to define precisely the rights and responsibilities associated with various managerial positions. But CPSU leaders were reluctant to prescribe proper behavior in situations when officials were forced to choose among contradictory or unclear directives. As a result, leading party, governmental, and economic officials often had to decide by "departmental" and "local" criteria whether their expectations of others and others' expectations of them were feasible and legitimate and what sanctions, if any, would be brought to bear under various circumstances.

Fifth, present-day Soviet citizens receive the media in private settings. In the Stalinist period, party activists read newspapers aloud to poorly educated or semiliterate groups, and collective discussion reinforced ideological homogeneity. Now, with virtually 100 percent literacy achieved and with modern telecommunications, people read newspapers, magazines, and journals privately and watch television and listen to radio at home in a family setting. Soviet theorists consider the privatization of the relationship between the consumer and the media to be an objective process in the modernization of communications. According to two Soviet analysts, "The development of satellite relaying, of multichannel systems of cable television, and of videotaping in the home will soon individualize the use of the television set, which today is a medium for group [family] consumption and will lead to the shaping of a new set of stable relations between television and its millions of consumers."[27]

Sixth, Brezhnev's strong support of television may have impeded traditional efforts to rear the "new Soviet man." Effective television programming capitalizes on the spontaneous potential of the medium, which makes censorship difficult.[28] Also, Soviet audiences (some demographic groups more than others) perceive television as an entertainment medium and resent its use for pedagogical purposes. Television is supplanting other available cultural forms, especially among Soviet urban workers.

Political-administrative Consequences. One important consequence of these internal and external developments is the undermining of the monopolistic access to information enjoyed by the top Soviet political elite. The priorities of a self-sufficient strategy of modernization are being supplemented by new socioeconomic demands, domestic and foreign, and by new technological means of pressing these demands, especially the worldwide revolution in telecommunications.

Power in contemporary Soviet society is integrally connected to the politics of information, and access to pertinent information is highly correlated with one's position in the party-state hierarchy. For example, four types of TASS bulletins are printed in different colors to signify the level of secrecy and the authority of its readers. Violet TASS contains material that can be reproduced in whole or in part. White TASS contains confidential data sent to the editors-in-chief of newspapers. Red TASS contains recent top-secret information that may contradict the lower-ranking TASS bulletins, and it is delivered by armed guard and placed in locked safes. Finally, TASS special bulletins contain reports from the Western press, as well as domestic news, and are issued only to the very highest party and government officials.

The growing significance of expertise, however, places mounting pressures on this hierarchical stratification. During the NTR the bureaucratic elites and citizenry must have wider access to information in order to formulate and implement effective policies. Policy-relevant information is more diffused in a maturing socialist system than it was in the industrializing state socialist system, and the mass media provide important channels for the circulation of such information. The proliferation of specialized knowledge and the central institutions' growing need to utilize this knowledge have led to a broadening of meaningful public communications. Specialized journals, for example, have become essential to the development and exchange of the expertise required to manage a complex society. As an anonymous commentator noted in 1978,

> In some of the societies of the Soviet type, conflict between the need to maintain the hierarchical order and to assure a minimum of non-authoritarian communication is today overcome in the following compromise. Even in the domain of the social sciences, the state apparatus intervenes only exceptionally in the choice of scientific paradigms

> by means of which it is hoped to resolve given problems; rather, it is the prohibition against raising particular problems that is used as a means for controlling research.[29]

In short, a wide variety of specialized journals has emerged to nurture the ideas and to disseminate the information useful to Soviet policy makers.

Policy groups have used the press to promote reformist, centrist, and conservative positions. For example, Evsei Liberman's and later Kosygin's economic reform proposals received a national hearing at least in part because A. M. Rumiantsev, editor-in-chief of *Kommunist*, provided valuable support to the innovative forces. In the early and mid-1960s, advocates of these proposals employed every form of available publicity, including newspaper articles, scientific reports, books, press interviews, and radio and television commentaries. Indeed, the debate between Soviet economic reformers and conservatives has been conducted in the mass media since the late Stalin period.[30] Especially frank criticisms and far-reaching proposals have appeared in the specialized Siberian journal, *Ekonomika i organizatsiia promyshlennogo proizvodstva*. But "discussions" were sometimes printed in the major newspapers and journals, and controversial views were often elaborated in books. In an unusually sharp exchange, the innovative recommendations of a leading management specialist were denounced by a conservative ministerial official in the pages of *Kommunist*. Particular arguments and claims were deemed "unconvincing," "erroneous," and "clearly confused," reflecting "ignorance of the way that the enterprises and ministries actually operate" and "a very poor understanding of our economic system." And then the coup de grace —the contention that Lenin would have rejected such Khrushchev-like recommendations:

> Lenin resolutely opposed the penchant for hare-brained organizational schemes and half-baked, ill-considered action, especially when it involved various reorganizations of management agencies or the hasty establishment of new institutions.
>
> What is needed at the present stage of industrial management is not to establish "superministerial agencies" but to strengthen and develop the existing branches, improve labor and production discipline, create a favorable moral and psychological climate among employees, and improve the work style and methods of the branch staffs—the ministries.[31]

Also, the environmental issue was brought into the Soviet political arena almost entirely by relatively independent forces, including ad hoc groups of citizens adversely affected by pollution, scientific experts, and investigative newspaper correspondents. As a Western commentator notes, "Local volunteers have used the media in their campaigns and may have thus had indirect influence on the Central Committee members through these vehicles. Scientists, too, have shown skill in using the media to publicize their environmental findings and perhaps bringing them to the attention of national policy makers."[32]

Furthermore, CPSU leaders periodically direct the editors of central newspapers and journals to encourage public discussion of key issues. Soon after becoming general secretary, for example, Brezhnev launched an open discussion about the economic activities of the party and the responsibilities of particular types of officials; Andropov encouraged economists and managers to propose diverse remedies for the ailing economy; and Chernenko precipitated a heated exchange among educators, production executives, and citizens that was followed by a reorganization of the nation's schools and a reemphasis on vocational training. To be sure, the party leadership initiates and terminates debates, restricts them largely to the printed media, often limits them to specialists, and forbids criticism of the fundamental characteristics of the Soviet polity. But Brezhnev, Andropov, and Chernenko often emphasized the complexity, interconnections, and dynamism of socioeconomic and scientific-technological developments, as well as the concomitant needs to tolerate ambiguities, to obtain more policy-relevant information, and to find partial solutions to ongoing problems. Even the highest CPSU officials do not claim to know the linkages among and the solutions to all contemporary challenges. They expect the media, however, to help elicit and disseminate such information.[33]

Although there is much ritualistic communication in the Soviet press, many important kinds of information are transmitted to many different audiences. The content of messages is determined largely by the intended audience and the purposes of the sender—not only by the sender's political position, analytical abilities, or access to classified data. Officials may write on the same topic for two different audiences and produce very dissimilar works. In the party press, a writer may carefully analyze specific administrative problems and explain the implications of recent Central Committee decrees. In works intended for mass circulation, the same writer may never even refer to a Central Committee decree. Problems meticulously analyzed in the specialized press suddenly become "unproblems." Ideal situations are described, not actual ones, and the distinction between desirable and existing conditions is consistently, if not deliberately, blurred.

Certain published Soviet documents shed light on the adaptive behavior of the CPSU's top policy-making bodies and of subgroups within regional party and government organizations. Articles that appear in the journals *Kommunist* and *Partiinaia zhizn'* are generally written by and for upper- and middle-level party officials; reports in *Propagandist* or *Agitator*, for example, are usually written by local party officials and rank-and-file Communists. Through these four journals alone, cadres at all levels of the CPSU continuously transmit information to one another.

Soviet public communications invariably transmit expectations that attempt to alter the beliefs and behavior of institutional, nonassociational, and amorphous groups in the polity and society. Published directives from the Central Committee and the Council of Ministers, as well as the materials of the party congresses, are of obvious importance. These sources communicate information about the immediate and long-range goals of CPSU leaders. Also, such sources reveal information about progress toward the achievement of desired goals—in actual fact and/or in the estimation of leading party officials. Moreover, a Central Committee decree forcefully transmits expectations. It often analyzes the administrative problems of a single province or city and issues detailed instructions to individual party bureaus. These authoritative policy pronouncements command, cajole, and persuade. Top party officials consider such directives binding on *all* CPSU organizations, even though a resolution may focus on the shortcomings of only one party committee.

Articles by provincial and local officials communicate information to various audiences too. Information is transmitted when cadres make simple statements of fact, send directives to lower-level party committees, and present "progress reports" to their superiors. Information is communicated through emphasis, omission, and tone. Members of various audiences quickly learn that certain themes contain particularly important cues and tend to provoke particularly revealing responses. Cadres also understand that "shortcomings" are frequently listed but rarely analyzed. In lieu of analysis, a CPSU official may briefly describe a single incident or type of situation and thereby communicate considerable information. The reader is invited to draw his or her own conclusions—a common practice in the Soviet press.

With the Soviet print media's shifting role and rising importance, the occupation of the journalist is becoming more important. Journalists have become a better educated and increasingly specialized group. Beginning in the 1950s two developments encouraged the professionalization of journalists: the formation of the Union of Journalists and the emergence of schools of journalism. By the mid-1960s more than twenty universities throughout the country prepared students for journalistic careers. The proportion of journalists with advanced journalistic training rose steadily. Graduates of journalism schools tend to work in fields such as science, international affairs, economics, and culture, whereas journalists who have graduated from higher party schools and other educational institutions often specialize in party propaganda and agricultural affairs. A journalism education is more and more a prerequisite for editorial positions, but party membership remains crucial. For instance, virtually all delegates to the congresses of the Union of Journalists are CPSU members.

At the core of the Soviet communication system is censorship, of which there are three basic types: the formal censorship of *Glavlit* and of the party's propaganda departments; editorial screening of deviant lines of thought; and self-censorship.

The process of formal censorship revolves around approval from the censor. After a newspaper, magazine, or book has been typeset, an editor reviews the proof copy. Then the copy goes to the censor located at the editorial office. The censor must review each page, sign it, and stamp it with his or her own number and the date. In practice, the censor rarely interferes with publication. According to Lev Lifshitz-Losev, "This is explained by the fact that editors' offices, as well as radio, television, movie, and theatre studios, all have veteran editors who know that any serious mistake which is noticed by a censor could have even more serious consequences on their careers. The censor, like a policeman on the beat, exists not to stop a crime, but to prevent it with his intimidating presence."[34]

The basis for exercising censorship is the "List of Materials and Information Forbidden for Open Publication in the Press." This list is printed each year with additions and changes, and every office connected with the media has a copy with a specific number stamped on it. Most of the items concern military and state secrets. But the list defines many broad categories, as well as specific items, requiring protection.

Zhores Medvedev maintains that since 1971 editors-in-chief of newspapers, heads of publishing houses, and censors specializing in particular disciplines may examine the personal files of authors, including classified information from the KGB.

> Dissident scientists working in minor research positions in the natural or technical sciences can get their papers published in academic journals, but their *books* are rejected or suffer endless delays in the process

> of consideration even when of very high standard. They are not able to publish more popular scientific articles in magazines with wide circulation. Publication of an experimental paper is considered normal, but the writing of a monograph or popular science book is regarded as a privilege of prominent and distinguished scientists. . . . This differentiation between categories strictly academic, scholarly, and general is achieved through different levels of blacklisting.[35]

Looser standards of censorship are applied by the regional and local press. It is not uncommon for a book or an article to be published in the provinces after being rejected in Moscow, Leningrad, or Kiev. Also, major and minor publishing houses occasionally ignore the censors' decisions or bypass the censors by printing limited-edition books.[36]

Many developments are modifying the impact of censorship on the Soviet population. News from abroad is limiting the party-state's ability to control information, and the USSR's participation in the world economy is reducing information closure. As Zhores Medvedev observes, "Soviet science is not only the instrument of development of Soviet economic and military power, it is also the main channel of knowledge about the outside world and cooperation with the outside world." But Soviet censorship prevents the scientific-technical intelligentsia from obtaining international information, and Soviet bureaucratism prevents some of the most able and active junior researchers from obtaining pertinent information promptly. Medvedev adds, "The secrecy of information is quickly being transformed into the *absence of reliable information*."[37]

Brezhnev and his colleagues recognized a few of the disadvantages of the dogmatic and centralized control of information. They perceived Soviet society to be more and more diversified and permeable to external influences, and they found it harder and harder to cope with, let alone to guide, socioeconomic and scientific-technological forces. Because the CPSU leadership recognized the importance of specialized elites, it increasingly used the media as a forum in which officials and experts could debate policy alternatives and alternative methods of implementing policies. Also, the media increasingly enabled average citizens to voice their opinions about policy outcomes at the local level.

The shift in the functions of the media was considerable. Khrushchev once told a group of journalists: "You are really always at hand to assist the party. As soon as any decision must be explained and carried out, we turn to you, and you, as the most faithful transmission belt, take the party's decision and carry it into the very midst of our people."[38] In contrast, the

Brezhnev administration quickly emphasized that the press was a multidimensional channel of communication. In 1967 a *Pravda* correspondent observed: "The provincial party committee's mail is not only an important means of studying the state of affairs in the localities and public opinion, but it is also a means of influencing this opinion, generalizing advanced experience, and eliminating various shortcomings."[39] The national and regional press continue to perform all of these functions.

Soviet analysts contend that the mass media must stimulate all workers and peasants to participate actively in resolving current domestic dilemmas. The press, radio, television, and letters from the citizenry are expected to criticize and recommend means of overcoming shortcomings, especially those of an administrative and socioeconomic nature. Soviet commentators carefully distinguish between party and state activities, a signal that public criticism of the CPSU as a whole and of its leading personnel will not be tolerated. Constructive criticism of regional and local bodies, including CPSU organizations, is permissible and is periodically encouraged, especially through the national, regional, and local soviets and their standing committees. Butenko observes:

> The concept of "power through the laborers themselves" certainly presupposes the broad utilization of the representative system. But significant changes are inherent in the implementation of such power. It is inevitably reorganized in such a way so as not only to provide information to the laborers about all vitally important issues of state policy, and not only to permit resolution of these issues with the assistance of representatives of the laborers, but also to consider and, most important, to utilize the opinions of the laborers *directly* in this effort. Hence, the roles of public opinion, of *open discussion of proposals*, and of *debate* about *urgent* issues pertaining to *all* spheres of social life are expanding. While the representative system is retained, the role of representative organs expands in comparison with that of executive organs, and the laborers' surveillance of the activities of executive state organs increases. . . .[40]

Moreover, CPSU leaders are counting on new forms of "intellectual technology" to improve decision making. Underscoring the significance of public opinion polling for scientifically grounded decisions, Shakhnazarov affirms:

> Today it is hard to establish goals in social policy without making a deep analysis of the social groups and a tentative, preliminary

> assessment of their possible reaction to various measures. . . . Effective propaganda depends to a very large extent on how much it serves as a two-way channel of communication, how much it plays the role not only of the helm but also the detector of public opinion. . . . Public opinion polls and other forms of sociological surveys make it possible to establish more accurately which questions are causing the greatest concern in a given social milieu at a particular time.[41]

In short, Soviet analysts affirm that the mass media play an important role in shaping and implementing policies. Two-way information flows, vertical and horizontal, must accurately reveal change and continuity in the material sphere as well as in education and culture.[42]

The increased flows of information between the party-state and the public are to aid primarily in the implementation, and secondarily in the creation, of effective policies. But the importance of the latter increased considerably in the Brezhnev years. As Safarov observes, "The value of public opinion under developed socialism consists in the fact that it participates in working out party policy and finding the most rational ways and means of carrying out the policy."[43] Safarov maintains that the elites and masses need more and better knowledge of many kinds. "The greater the totality of information communicated to society, the higher is the degree to which public opinion is informed and competent in matters of monitoring the administration, the ministries, and the executive committees."[44]

The press has long been a major feedback channel in Soviet society, and citizens sent letters to central and local newspapers even under Stalin. But unprecedented today is the quantity of letters, the diversity of the questions raised, and the institutionalization of the official response to unsolicited correspondence. Each of the central newspapers receives over a half million letters annually, and the CPSU Central Committee receives almost a half million additional letters.[45] To cope with this deluge, a letters department was created in the Central Committee apparatus in 1978. More regularized procedures were established for channeling letters on various subjects to various organizational units, and time limits were set for detailed responses from appropriate officials. Not surprisingly, such letter writing and administrative responsiveness have been heralded as manifestations of democratization in a developed socialist society.

An illustration of how letters are used is the *Literaturnaia gazeta* "If I Were Director" department, a series of readers' proposals on improving various institutions and activities. From 1974 to 1977 about 8,000 readers contributed 17,000 suggestions. "The newspaper assembled 800 of the proposals in the form of a bulletin. The USSR Council of Ministers studied this bulletin and then instructed the heads of many ministries, agencies, and city soviets to comment and act on the suggestions."[46]

The press, and *Literaturnaia gazeta* in particular, has helped to vent social problems. Readers' opinions are solicited either by questionnaire or by letters. The discussions concern socioeconomic questions, the answers to which are not presumed to be already known. These exchanges are rarely efforts to inculcate fully developed and "correct" ideas but are relatively open-ended examinations of such issues as work, leisure, education, public services, the family, and so on.[47] To be sure, the authorities monitor all of these discussions and manipulate others. But real issues are raised, and pertinent information about *alternative* policies is communicated to and evaluated by a large audience.

Furthermore, "video" communication between the political leadership and the masses is increasing, and this two-way exchange is a double-edged sword. Not only does television enable CPSU and government leaders to "reach" a wider audience, but it enables the citizenry to observe and evaluate them. Competence and charisma, bureaucratic arrogance and ineptitude, and human foibles and frailties are communicated far beyond the walls of the Kremlin. Brezhnev, Andropov, and Chernenko were all seen and heard in enfeebled states, and in a by no means trivial example, Brezhnev was once observed picking his nose for some time on national television. For another example, Omsk television presents a wildlife series that "does not limit itself to descriptions of nature's beauties and rare animals but takes a socially active position." This show has demonstrated that the local bureaucracies often oppose environmentalists' concerns. According to M. Odinets, "The television camera introduced viewers to the people with whose connivance nature was being damaged. And here what was perhaps the strongest aspect of the program came out. *All the nuances of human behavior show up on the television screen*. It is immediately apparent who is dissembling, who is dodging, who is simply unprepared to answer a question, and who is speaking the bitter truth, but nonetheless the truth."[48]

Still another example of video communication is the monthly Ukrainian television program, "Press Club," which is designed to inform the residents of Kiev about current problems in their city. Heads of municipal services are invited to participate, and viewers are encouraged to call the studio and ask questions. For some time almost every show was disappointing to its producers, "due primarily to the unwillingness of certain officials to take part in answering the working people's questions." And when city officials occasionally demonstrated their ignorance of local conditions, "embarrassment reigned in the studio."[49]

Brezhnev's focus on problem situations and his use of the press for conflict resolution may be viewed as responses to advanced modernization or a "rationality crisis" in developed socialist society. In order to formulate feasible policies and to carry them out under rapidly changing conditions, mature socialism requires greater freedom of inquiry and debate than traditional state socialism allowed. The role of the media, particularly of the press, has thus been altered and enhanced. A modicum of "negative" freedom has become increasingly necessary to strengthen the central institutions' will and capability to exercise "positive" freedom for the common good.

Socioeconomic Consequences. We will now elaborate on six dimensions of the adjusted and augmented role of the Soviet media in the era of the NTR and some of their socioeconomic effects.

First, the Soviet media are important in the process of economic modernization. The economic reforms of 1965, 1973, and 1979 deemphasized the enterprises' traditional performance criteria, such as gross production totals, and emphasized new criteria, such as planned deliveries. The press made possible the exposition and dissemination of innovative ideas. Competition among major schools of thought concerning the reforms took place in the press. New forums such as the "discussion" section in *Ekonomicheskaia gazeta* addressed production issues. According to L. I. Lopatnikov, the press must "above all deal with economic methods of influence on the performance of enterprises."[50]

Soviet analysts underscore several responsibilities of the press in the economic sphere. The press is to disseminate general knowledge about the factors promoting growth and productivity because public education is thought to have a major indirect effect on economic performance. Also, the press is to publicize effective organizational methods and technologies and to encourage experimentation in economic management. For example, the party decided in 1977 to upgrade the training of managers after a two-year discussion in *Literaturnaia gazeta*. Furthermore, the press is to criticize economic shortcomings. Rather than "superficially" evaluate economic performance in terms of compliance with traditional plan indicators, journalists must scrutinize economic results and assess them on the basis of an increasingly complicated set of criteria. According to V. Seliunin, the press must make "independent analyses" of the performance of enterprises and play "a big role in the reordering of values." The press must extend its participation in economic modernization to include criticism of its own analyses of economic progress. All too often, however, "when we journalists write about lagging enterprises, we seem to forget all about [economic reform] and go on in the same old way."[51]

Second, the gradual shift from a manufacturing to a service economy affects the Soviet media in several ways. The media are expected to emphasize that this shift is one of the "historical" advantages of socialism over capitalism. Instead of conceiving of the growing service sector as a general attribute of advanced industrialization, commentators emphasize the "humane" quality of socialist economic growth and the valuable public services the economy provides. The press is to play a major role in persuading the population that only socialist systems can produce scientific-technological *and* socioeconomic progress.

The press is also involved in evaluating the actual delivery of services. Poor services are not linked to general systemic failures. Rather, the shortcomings of specific officials, programs, and organizations are criticized. For example, a *Literaturnaia gazeta* reporter sharply attacked the Ministry of Railroads in 1978 for evading inquiries concerning readers' complaints about railroad performance. Undaunted by hostile and indifferent government attitudes toward the press and supported by his superiors no doubt, the correspondent urged the Minister of Railroads to reject "an opinion I heard repeatedly from prominent officials of the ministry you head—an opinion about the functions of *Literaturnaia gazeta*—that our newspaper has no business concerning itself with transportation questions. . . ." Buttressing this claim with an implicit threat, the commentator noted: "We are getting a flood of letters now about the railroads . . . and readers accuse the newspaper and the ministry of having failed as yet to offer a thorough explanation of the reasons for slowed train movement."[52]

Moreover, the growing Soviet media are a key part of the trend toward a service economy and are shaping the use of people's leisure time and perhaps job time as well. Soviet sociological surveys often find that the three most popular forms of leisure activity are listening to the radio, reading newspapers, and watching television. The NTR is increasing the population's free time through modest though declining gains in labor productivity. Also, the NTR is extending the reach of the media through the dramatic rise in the production of television sets and in the expansion of broadcasting capabilities. Notwithstanding the paucity of commercials, Soviet television programming may well stimulate people's interest in desirable life-styles and consumer goods that can be obtained only by skipping work or by laboring chiefly on lucrative part-time jobs.

Soviet analysts acknowledge that the quality and differentiation of the media must be improved in order to fulfill the needs of a service-oriented population. S. V. Tsukasov, managing editor of *Pravda*, candidly reflected on a major sociological study of "the dynamics of the reader-newspaper relationship."

> [*Pravda*] readers display heightened interest in important topics: economic matters, international affairs, the upbringing of young people, and science.
>
> It appears also that there is a certain gap between the readers' "demand" and what the newspaper supplies them. If one lists the sectors on which our work ought to be improved from the standpoint of readers, they turn out to be chiefly the key problems on which the party is focusing public attention. The readers' interests are linked with fuller treatment of party affairs, particularly questions of the communists' activeness and responsibility and the style of work of party committees; ways of improving production and management, the communist ethic; and the problems of peaceful coexistence and ideological conflict in the international arena. The reader wants more information, more diversity of writing, and higher journalistic quality.[53]

Third, the worldwide communication revolution is having a mounting impact on Soviet society. The mass media in the United States and the Soviet Union have as their common denominator the sheer pervasiveness of messages. While U.S. messages are overtly commercial and Soviet messages overtly political, the citizens of both states face a similar predicament: how to shield themselves from the bombardment of pressures contained in public communications. Both intellectual and emotional detachment are difficult. What appears in the media is "real"; what does not has an aura of privation or isolation. New forms of alienation may be stemming from people's inability to detach themselves from the outpourings of the communication industries and to develop their own perspectives on what is important to themselves and to society.

The Soviet–American competition on earth and in space is an important stimulus to technological progress in telecommunications, which in turn is having socioeconomic consequences throughout the USSR. The state of Soviet communications in the mid-1980s is graphically portrayed in table 1.[54] The Molniia satellites (I, II, III) are obsolete. They are slow-moving and require linkages with several orbital stations. Also, the Molniia capacity for television is overloaded because the system must relay programs from Moscow to the whole of Siberia and the Far East. The recep-

Table 1 Summary of Soviet Satellites for Civilian Telecommunications

Group (general characteristics)	Type	Number 12-31-83	Orbital Life
MOLNIIA ("Lightning")			
—since 1965, one launched every 110 days;	I	55	15–20 yrs
—5 Molniia I's and 3 Molniia III's in range of each of 100 Orbita ground stations at any one time (though not all necessarily operative);	II	19	20 yrs
	III	19	12 yrs
—Elliptical orbit: perigee = ~600km; apogee = ~39,800km			
—Standard inclination 63 degrees over the equator.			
MOLNIIA I-S	I-S	1	—
—first operational Soviet geosynchronous satellite;			
—launched 29 July 1974;			
—test vehicle for the Statsionar series.			
STATSIONAR			
—all geosynchronous satellites;			
—each Statsionar has an operational life of approximately two years.			
RADUGA ("Rainbow")	I	13	—
EKRAN ("Screen")	II	11	—
GORIZONT ("Horizon")	III	7	—
LUTCH			
—launched March 1982.	—	1	—

tion of these programs is poor or nonexistent in large parts of the nation outside of European Russia. Mickiewicz reports, "Some three-quarters of the Soviet population live in 'zones of secure reception,' and here some 85 to 90 percent of the families have television sets."[55]

To expand telecommunication capacity, Radouga satellites have been introduced. They are launched into a geosynchronous orbit, which means that the Radougas, unlike the Molniias, have a fixed position in the sky. The immobility of the Radougas eliminates the need for a costly aerial system that must be constantly moved from one position to another while orbiting. But, in order to maintain a fixed position, the new satellites must be placed in orbit immediately above the equator, which makes it difficult for them to reach Eastern Siberia. Improved transmission to this region is the aim of the Ekran satellite launched in 1976. It broadcasts directly to television sets and avoids the use of intermediary receivers. Hence,

Frequency	Aim	Remarks
800–1000 MHz 6.2 GHz	To provide a satellite telecommunications system internally and with the COMECON nations.	Though perhaps dated equipment, Molniia's have reliably formed "a major part of the Soviet Union's communications system."
6.2 GHz	To provide communications links during manned Soyuz flights.	Molniia II —last launched Feb. 1977; —used as "hotline stations" between the US and USSR.
6.2 GHz		
3.4–3.8 GHz	To provide for domestic telecommunications.	Aim is to maintain one operating and back-up satellite at each Statsionar location—with the Ekran, for example, satellite breakdown could mean "complete loss of domestic TV services."
702–726 MHz	To provide television broadcasts directly through aerials of very small diameter.	
—	To provide for international relays such as the Olympic games.	
11, 20, 30 GHz	To evaluate super-high frequencies for future communications networks.	An Intercosmos project with several Eastern European nations participating.

urce: *Jane's Spaceflight Directory*, 1984. Table provided by Adam Siegel.

the Radougas and Ekran are designed to supplement the Molniias.

A telecommunication system performs two basic functions. On the one hand, it can increase the number of people watching television. Over 80 percent of the Soviet population owned television sets in the early 1980s, and telecommunications are soon to reach most of the remaining people, who primarily live in Eastern Siberia and the Far East. According to a 1976 *Pravda* editorial, "It has been determined economically feasible to transmit television programs there via communication satellites to receiving stations that can be installed in small towns and villages. Such systems, together with a network of radio relay and cable lines, will enable us to push back the boundaries of television broadcasting."[56] Eastern Siberia and the Far East, regions vital to the economic future of the USSR, must attract and keep workers, and television is an important mode of entertainment on a harsh frontier. Also, telecommunications speed the news to remote areas. A page of *Pravda* can be transmitted to Far Eastern printing presses in only three minutes by satellite, rather than the twenty-two minutes required by cable or radio.[57]

On the other hand, modern telecommunications can broadcast programs from more than one television channel to even the most isolated areas of the USSR. Numerous channels in Western Russia have already expanded consumer choice. Also, Mickiewicz found that "the higher density of sets in urban areas does not mean that urban residents watch more television. In fact, the reverse is true. There are 1 1/2 times *fewer* television sets in the countryside than in the city, but television-watching time in the countryside *exceeds* that in the city by 1 1/5 times. Among the Soviet rural population, television is unquestionably the most popular and widespread form of mass communication."[58] These trends will probably continue as the quantity and quality of telecommunications improve. Although the USSR has remained aloof from Western telecommunication systems, it has developed an independent capacity and established a telecommunication treaty with its allies. *Intersputnik* was designed to rival *Intelsat*, so as to provide the Soviet bloc with regional programs.

Of great importance to all media is computerization, a significant trend in the United States. But civilian telecommunications are a low-priority sphere for computerization in the USSR. They will probably be built up slowly and primarily with foreign trade. Although the 1980 Olympics provided a stimulus to upgrade Soviet computerized telecommunications, the Jimmy Carter administration vetoed a computer deal with TASS in 1978. The cancelled text-editing system was chiefly designed to help TASS cover the Moscow Olympics, and Soviet officials maintained that the extensive news coverage required computers to file correspondents' reports and transmit them abroad.

Soviet analysts understand that each element in the communication industry—books, journals, newspapers, radio, and television—possesses specific strengths and weaknesses. A differentiated communication policy is to be perfected. According to I. G. Petrov and A. S. Seregin,

> The press engages in detailed, particularized treatment of a problem, description of facts and events, theoretical explanation, interpretation of facts, and analysis of their interconnections. The radio presents timely, vivid, easily understood messages about facts, cleansed of unnecessary details and perhaps even somewhat incomplete (to stimulate recourse to other information sources). Television provides on-the-spot reporting of events (creating the direct-participant effect),

> as well as visual demonstration of events and their most important components and most vivid episodes, expressing the essence of events.[59]

Hence, a modern communication system is thought to be highly diversified, maximizing the effectiveness of propaganda messages and the fulfillment of consumer needs. Soviet sociologists are striving to create "a system for selecting particular information channels, with consideration for the psychology of their perception and assimilation by society."[60]

Fourth, the growing significance of scientific and technological knowledge has begun to affect the theory and practice of journalism in the USSR.

Soviet analysts affirm that journalistic education must be considerably improved. A party decree in 1975 called on journalists to study the complex and interrelated problems of scientific-technological, socioeconomic, and political-administrative development. New programs at Moscow State University were established to train economic journalists and to examine "urgent questions of the theory and practice of the press, television, and radio."[61] According to *Pravda*'s Tsukasov, the education of journalists has improved but "remains well below the requisite level. The Achilles heel of journalistic training continues to be lack of professionalism and especially insufficient on-the-job practice."[62]

The nature and level of television journalism are of special concern. A Soviet analyst declares, "The journalist has yet to assume his proper place on the screen and very often gives way to the announcer."[63] In order to develop the unique potential of television, journalists must "effectively and vigorously" master such broadcasting forms as reportage, interviewing, and commentary.

Soviet spokespersons contend that scientific discoveries and technological innovations must help to modernize contemporary journalism. Drawing on systemized knowledge, such as public opinion research, journalists are to develop a science of media management. Tsukasov recommends that a newspaper's long-range plans be issue-oriented, with key short-term subtopics assigned to specially trained staff members. Also, he urges that mathematical models and computers be used to facilitate quantitative evaluations of materials under consideration for publication.

> *Computer equipment* is required to study mailbag content and its topics by geographical regions, to take the pulse of public opinion, discover its dynamics, and know which problems demand attention. The data we obtain should guide us in adjusting editorial plans and

> campaigns. The data should also be brought to the attention of interested party and governmental agencies.
>
> Unfortunately, *even central papers now lack the equipment* to conduct extensive structural study of their mail or to analyze criteria of effectiveness such as the frequency of their treatment of key problems, the content of the news flow, the social characteristics of the newspaper's "heroes" and contributors, the direction and social utility of newspaper criticism, the geographical spread of items carried, and so on.[64]

By 1980, as Stephen White recounts, this situation had changed.

> The letters department in *Pravda*, 70 strong, is apparently the paper's largest. Every letter the paper receives is first of all given a control card, on which are entered the sex, apparent age and occupation of the writer, and a brief summary of the contents. A *computer* is then used to calculate a breakdown of the themes dealt with in the letters and the parts of the country from which they come, and every month a printout of this information is made available to the deputy head of the department. Regular thematic reviews of letters received are also prepared, classified by the part of the country and the institution concerned, and then passed on to the appropriate authorities.[65]

In short, Tsukasov's calls for computerization were heeded. Additional items on his agenda for the technological modernization of journalism included central data banks, as well as the automation of news reference services, copy editing, make-up, and certain administrative functions.[66]

The growing significance of science underscores the importance of reporting scientific and technological advances in the press. The journalist specializing in the social or natural sciences must not merely popularize scientific achievements but must nurture the climate for scientific-technological and socioeconomic progress, according to *Izvestiia*'s science and technology editor.[67] Also, the journalist is to be an interlocutor with the scientist, exploring possible areas and methods of mutual assistance. In exceptional cases, the journalist and scientist are one and the same person. Reviewing two books by Afanas'ev, the Soviet commentators declare:

> It is especially gratifying that the books are written by a scientist, academician, and at the same time editor-in-chief of our most important newspaper, a journalist who appears in *Pravda* and other organs of the press, as well as on radio and television, with noteworthy materials.

> Doesn't this mean that in our day journalism and the press are inseparable from science? The demands of a scientific attitude, objectivity, and accuracy are the Leninist demands on the press. But that is just what science is. It seems that the source of many of the ideas and facts contained in the reviewed books is the newspaper where the author works. Hence, too, the clear, intelligible language of the books, even, at times, publicist language.
>
> The alliance of journalism and science seems very fruitful to us. And evidently it is no accident that the number of scientists who work in the press is growing, and that scientists are not avoiding journalism.[68]

The media, especially journals for well-educated audiences, are viewed as important means of integrating expertise and of reducing the isolation of specialists in related fields. V. Druin observes, "Contacts with journalists help specialists in various sciences to find the general language necessary to communicate with one another and with the public."[69] For examle, *Voprosy filosofii* has developed and disseminated new ideas about the science-society relationship. In fact, Soviet leaders rebuked the journal for playing this role *too well*, and, in the mid-1970s, more conservatives were added to its editorial board. Critics noted, "The editorial collegium has published many articles *by natural scientists* on problems having a philosophical aspect. In the future, however, the editors should play *a more active role* in defining the philosophical problems of science that should be covered in such articles."[70]

Sixth, the Soviet mass media are having various influences on culture. According to Soviet surveys, workers and peasants spend more time watching television than the *combined* time they spend reading newspapers and books and going to the movies and theater. The smaller one's city or the more isolated one's rural area, the fewer the cultural outlets. With a decline in cultural competition, the dominance of television increases. In contrast, the highly educated residents of large cities are most likely to be consumers of the printed media, the cinema, and the performing arts.

Soviet analysts are studying the media's cultural effects, such as the debasement of language and the addiction of children to television. But perhaps the most significant and subtle impact of television is on mass political culture, particularly attitudes toward authority. We have already cited examples of video communication between the CPSU leadership and the adult population. For another example, children's exposure to television undermines the teacher's authority in the classroom. As a Soviet commentator observes, "If pupils once accepted unquestioningly everything the teacher said, now they challenge him constantly on the basis of movie or TV productions of the literary works they are discussing. And often they stymie him."[71]

In addition, the consumerist tendencies of Soviet culture conflict with the public values promoted by the mass media. Soviet theorists maintain that party propaganda must combat "commodity fetishism," ennoble one's pursuits, and instill a sense of personal responsibility for all aspects of collective well-being. Soviet society is to curb "degenerative" Western influences by enhancing law and order and by reducing social anomie and alienation. Consumerism, in particular, is thought to be a manifestation of capitalism, and the media have been enlisted in the struggle against this "petty-bourgeois psychology." In lengthy discussions in *Literaturnaia gazeta* and elsewhere, Soviet correspondents and readers invariably identify consumerism as one of the chief sources of Western influence penetrating the USSR through films, books, and radio broadcasts.

Mickiewicz concludes: "The Russian public, when examined scientifically, turns out to be much less homogeneous, monolithic, and malleable than the Soviets (and Western observers who were, perhaps, persuaded by the Stalinist theories of their communication efficacy) had thought. . . . The real question for the leadership is no longer whether public opinion should be listened to, but how much."[72]

Developmental Options and the Dissident Movement

The two basic roles of the Soviet media have been discussed above. On the one hand, the media are instruments of ideologically motivated political power, advocates of official policy, pedagogues of the population, and promoters of positive freedom. On the other hand, the media are arenas in which "nonantagonistic contradictions" are expressed, forums for different viewpoints on political-administrative problems, participants in the resolution of such contradictions and problems, and facilitators of negative freedom.

The media have become an important channel of conflict between the traditional strategy of industrialization and the advanced modernization strategy. State socialism used the media to implement predetermined positions. Developed socialism underscores the legitimacy of freedom of debate on many issues, and the media—especially the press—enable citizens to exercise this freedom within the parameters set by the political authorities.

The conflict between the traditional and developed socialist strategies is largely about the nature of scientific knowledge and the role of experts in society. State socialism conceptualizes science primarily in terms of dialectical and historical materialism, which posits the Marxist-Leninist "laws" of socioeconomic development and defines freedom as "the recognition of necessity." In contrast, some Soviet analysts conceptualize science primarily as organized information, which derives from the cumulative empirical study of human and physical behavior. The Brezhnev and subsequent administrations have attempted to balance these two approaches. Contemporary science is thus becoming a new and disputed kind of "guide to action." Soviet conservatives stress that science and technology are valuable instruments to accelerate irreversible trends. Soviet reformers emphasize that science and technology are critical components in *defining* as well as meeting the challenges of mature socialism.

Leading CPSU officials agree, however, that significant information flows must be regularized and controlled. In an interview on the American television program *60 Minutes*, Afanas'ev was asked: "The fifty-million people who read *Pravda* every day, they do not hear how much foreign aid you give to Cuba, how many casualties you have had in Afghanistan, how many tons of wheat you buy from the United States. Why not? Is that not news?" Afanas'ev replied: "Well, this is news, but we consider that our Soviet reader can live without it. I think he would survive even if he wouldn't know how much and what we have sent to Cuba and how we did it. This is not the news which is absolutely vital for human survival."[73]

Also, the post-Stalin communication system has curbed public discussion of the overall meaning of trends, especially the interrelationships among scientific-technological, socioeconomic, and political-administrative developments. To be sure, the Soviet media have transmitted increased amounts of specialized information from top to bottom, from bottom to top, and laterally throughout the polity. But the party leadership has legitimized only the discussion of concrete issues, not their relationships to essential features of the Soviet system or their implications for changes in the system as a whole.

The Soviet scientific-technical intelligentsia has been given considerable leeway in producing further expertise and in periodic public discussions about the political-administrative applications of its expertise. Specialists, however, cannot freely debate issues broadly affecting the evolution of developed socialist society. As Zhores Medvedev remarked in 1978, "Research workers are now given much more *scientific* freedom, while political freedom remains extremely limited."[74]

The dissident movement in the USSR, which draws heavily from both the scientific-technical and cultural intelligentsias, rejects such compartmentalization of expertise. Virtually all dissidents, other than the neo-Stalinists, favor more open communications. This preference was reflected not only in the petitions of dissidents but also in their expansion of the "self-publishing" (*samizdat*) movement during the 1960s and 1970s. *Samizdat* directly challenges the principle and practice of censorship, providing alternative facts and interpretations to counter party-state control of the media and self-censorship. Whereas radical Western journalism emphasizes analysis and prescription, radical Soviet journalism focuses on description and objectivity. Hence, the underground *Chronicle of Current Events* eschews editorializing and concentrates on facts rather than interpretation. Joshua Rubenstein affirms:

> The principal goal of the *Chronicle of Current Events* has been to challenge the regime's control of information. It therefore has been eager to report on a wide range of nonconformist activities, involving everyone from anti-Stalinist Marxists to liberal reformers and evangelical Christians. But the *Chronicle* avoids commenting on its own reports, wishing to stand apart from the ideological tensions that arise among the dissidents themselves. Consequently, it is not a useful source of information for the debates—on detente, emigration, or the future of Russia—that have shaken the dissident community. . . . At the same time, by not commenting on these and other issues, the *Chronicle* gained the confidence of diverse groups. For in a society where genuine information is precious and tolerance is rare, the *Chronicle* has tried hard to avoid compromising the integrity of its reports with editorial comment.[75]

The expression of diverse interests and opinions, as well as the uses of expertise, are ongoing sources of tension in the USSR. An example of Brezhnev's efforts to prevent specialists from raising issues outside of their field is Zhores Medvedev's incarceration in a mental hospital. This case reveals not only that psychiatry was used to suppress nonconformist ideas, but also that scientists and technicians have been forcefully prevented from examining the relationship between knowledge and the purposes it may serve. Medvedev was declared to have "incipient schizophrenia" accompanied by "paranoid delusions of reforming society." The symptoms of the "illness" included a "split personality, *expressed in the need to combine scientific work in his field with publicist activities*; an overestimation of his own personality, a deterioration in recent years of the quality of his scientific work, an exaggerated attention to detail in his publicist writing, lack of a sense of reality, poor adaptation to the social environment."[76]

Medvedev's psychiatrist asserted, in effect, that his "patient" could be considered mentally healthy only if he refrained from commenting on political and social affairs. Zhores' twin brother, Roy, asked the psychiatrist: "But what right do you have to pronounce on works outside your own field and to ignore the opinion of people much more competent than you to judge the problems dealt with in Zhores's book?"[77] Although this argument had no impact on the treatment of the accused, it exposes the highly selective or arbitrary nature of the CPSU's principle that experts must not break the confining chains of their specialized research interests. As many Soviet dissidents know from first-hand experience, interdisciplinary study is stifled when bureaucrats forbid intellectuals to transcend analytical divisions or when intellectuals protect themselves against bureaucratic interference by claiming competence only within a narrowly defined specialty. Legitimacy is thereby denied to linkages among spheres of knowledge and to investigations of the relationships between knowledge and the interests it serves.

Liberal and democratic socialist dissidents strongly advocate the right to engage in "publicist" activities outside of one's own field. For Andrei Sakharov, the most prominent liberal dissident and Nobel laureate, exchanges of information among the Soviet bureaucratic elites, between the elites and citizenry, and between the USSR and other countries must be dramatically increased. Restrictions on discussion and data deprive "the middle-rank leaders of both rights and information, turning them into passive executors of instructions, into simple functionaries. The leaders of the higher echelons have information passed up to them which is too incomplete and too polished, and, moreover, they are deprived of the possibility of making effective use of the powers at their disposal."[78] Neo-Stalinists justify intellectual censorship on the grounds of preventing the "penetration of foreign ideology" and the "undermining of the foundations" of the Soviet state. But Sakharov declares that intellectual freedom is essential to creativity: "Fresh and deep ideas . . . can arise only in discussion, in the face of objections, only if there is a potential possibility of expressing not only true, but also dubious ideas."[79]

Roy Medvedev, at first a radical Marxist-Leninist and later a democratic socialist dissident, contends: "Almost all discussions of the problems of socialist democracy center around the question of the freedom of information, which includes freedom of speech and the press, freedom for scientific research and artistic expression."[80] He maintains that much freer flows of information are needed to establish "genuine" democratic centralism and to eradicate the ingrained "bureaucratic centralism" of the Stalin era. According to Medvedev, "The present system of bureaucratic control over information results in ignorance not only at the bottom but also at the top." Medvedev declares, "If the Central Committee were transformed into a truly political body linked with the masses and sensitive to their moods and wishes, the party would be able to initiate a *dialogue* with the dissidents, i.e., with all the various new trends of social and political thought in the country. It would mean a confrontation of political ideas without recourse to any kind of administrative prohibitions or repressive measures."[81]

Such liberal and democratic socialist tendencies were far to the left of mainstream Soviet political theory and practice in the Brezhnev years and thereafter. The centrist position has been a byproduct of the conflict between the traditional Soviet industrialization strategy and the advanced modernization strategy and of the uneasy coexistence between their respective communication systems. Domestic political competition and cleavages are likely to continue as the USSR becomes more permeable to external influences. Soviet reformers stress and Soviet conservatives question that global information flows are essential to socioeconomic progress in the USSR, especially scientific-technical data and market forces that spur economic growth and productivity. But the outlooks of leading CPSU officials have many more similarities than differences, especially vis-à-vis pro-Western dissidents.

Soviet analysts recognize that the NTR has sparked an "information explosion" within and among nations. This is attributed in part to a "dramatic and unparalleled expansion of communication technology" and to an "expansion of international economic, commercial, financial, scientific, technical, cultural, and other links."[82] Indeed, many regions of industrialized nations are becoming increasingly sensitive and vulnerable to pressures, ideas, and information from other regions and from abroad. Also, the leaders of industrialized nations are striving to improve the transmission and processing of information because effective decisions in virtually all issue areas depend on a continuous flow of accurate and timely data from domestic and international environments. According to Soviet observers, this is an inevitable process accelerated by the NTR.[83]

Soviet spokespersons sharply distinguish between different types of information. V. Korobeinikov declares, "Information in the form of scientific and technological data is connected with ideology only indirectly," but "a substantial part of [other] information . . . directly expresses class interests and is inseparably linked with the waging of the ideological struggle."[84] The traditional Soviet communication system underscores objective limits to cooperation and moves toward informational closure to protect socialist ideology from capitalist influence.[85] The communication system of mature socialism underscores international cooperation as well as the USSR's ability to compete in the clash of ideologies and interests.[86] Briefly stated, international conflict and cooperation are thought to be developing concurrently while continuously influencing one another. The perceived challenge to the USSR in international politics, and especially in knowledge-intensive fields, is to balance conflict and cooperation.[87]

The Soviet liberal and democratic socialist tendencies, however, reject such a balance and affirm the objective need for much greater cooperation, especially worldwide informational exchanges. Roy Medvedev exhorts: "Thus, whatever the inclinations of those in charge of ideology, we must learn to live in a more open world where there will be a comparatively free circulation of ideas and information. Clearly, this calls for a real transformation of our whole system of ideological work, our whole system of propaganda."[88]

The Soviet communication system became an international issue in the controversy surrounding the Helsinki accords signed in 1975. In a nonbinding and legally ambiguous declaration of intent, the signatories at Helsinki agreed to:

> *Make it their aim* to facilitate the freer and wider dissemination of information of all kinds, to encourage cooperation in the field of information and the exchange of information with other countries, and to improve the conditions under which journalists from one participating State exercise their profession in another participating State. . . .
>
> *Make it their aim* to facilitate freer movement and contacts, individually and collectively, whether privately or officially, among persons, institutions, and organizations of the participating States, and to contribute to the solution of the humanitarian problems that arise in that connection.[89]

The Brezhnev administration interpreted these pronouncements to be an extension of the entire thrust of the document—namely, to encourage international cooperation in the fields of economics, science, technology, education, culture, and the environment. In contrast, many Westerners viewed the same pronouncements as support for freer flows of information and human rights *within* the Soviet Union and for closer contacts between Western journalists and Soviet citizens.

Many Soviet dissidents also applied the Helsinki principles to domestic conditions. Responding to the CPSU leadership's public commitment to the "freer flow" concept, the dissident movement formed Helsinki "watch groups" in 1976. These groups monitored Soviet compliance with the Helsinki agreement and circulated their findings through *samizdat*. By early 1977 several reports identified Soviet violations and were communicated through foreign journalists in the USSR to the Western press.

The Brezhnev regime responded by crushing the Helsinki watch groups. The trial of Anatolii Shcharanskii was the most visible suppression of the monitoring effort. Party leaders found particularly objectionable the dissidents' implicit linkage between freer exchange of information among nations and greater freedom of association within the USSR. The increased dissemination of information was not to spawn associational interest groups. In addition, the CPSU leadership was determined to break the links between foreign correspondents in the USSR and dissidents. As Robert Sharlet noted in 1977, "Unlike previous KGB campaigns, which primarily sought to isolate or punish key activists, the thrust of the latest campaign has been to disrupt the communication channel that runs from the dissidents through the Western press corps in Moscow to American and European media and back into the Soviet Union via British and American Russian-language broadcasts."[90]

The Belgrade meeting of 1977, a follow-up to the Helsinki conference, unsuccessfully attempted to resolve the controversy concerning the freer flow of information and human liberties. The West, especially the U.S. delegation, provided a detailed critique of human rights violations in the USSR. The Soviet representatives responded by citing human rights violations in Western countries. Also, Soviet spokespersons vigorously defended the principle of national sovereignty in the information field. According to V. Chkhikvadze, Western anticommunist forces are "trying to take advantage of the expanded ties between East and West to promulgate alien bourgeois ideas in the socialist countries, thus warping the socialist public consciousness and undermining the people's ideological convictions."[91]

Conflict among nations about the exchange of information, as well as conflict in the USSR about conservative, reformist, and dissident aims and activities, persist. International disputes about information intensified after the Soviet Union proclaimed itself a developed socialist society and interacted more and more with capitalist nations. Domestic disputes about information also intensified in the USSR during the Brezhnev years. The traditional industrialization strategy had enhanced Stalin's authority and capability to formulate national policies with little participation by specialized elites. But a mounting information and decision overload and an expanding international division of labor undermined the authority and capabilities of Stalin's autocracy and of the rejuvenated party under Khrushchev. The subsequent collective leaderships mixed the traditional and advanced modernization strategies, underscoring the importance of expertise in decision making and implementation and encouraging officials and specialists to discuss pertinent issues in private and in the mass and specialized media. The implications of such expertise and issues for the restructuring of the polity, however, were not subjects of legitimate discourse.

Only the liberal and democratic socialist dissidents stressed the critical importance of unfettered dialogue between the rulers and the ruled. Roy Medvedev contends that the Brezhnev administration concealed a vast quantity of statistical data about "shortcomings, problems, and mistakes, [making] it impossible to mobilize the help of millions of people to overcome our failings."[92] An old dissident and former Bolshevik, S. N. Smirnov, observes:

> Stalin put Soviet statistics under lock and key. There was a certain rationale for this in the past. But the present situation is entirely different. There are no longer any forces in the world who could risk a military campaign against our country. There is nobody to be feared and no reason to hide our shortcomings or our statistics. But the lock is still there. And apparently the reason for this is no longer a fear of external enemies but misgivings that our own people, once they become familiar with certain statistical data, will begin to have doubts about the wisdom of certain past and present leaders of our country.[93]

In a word, the Soviet dissident movement calls for systemic change. Dissidents have intensified international and domestic conflict about information flows, and they have probably weakened the Soviet political system's considerable legitimacy at home and have definitely reduced its less-than-considerable appeal abroad.

The Soviet Polity and Dissent

To question the *wisdom* of your government's policies is one thing. To question the *fairness* of these policies is more significant. And to question the *authority* or *right* of your government to engage in certain kinds of activity and to pursue specific objectives is more significant still. The citizen who doubts the wisdom of public policies questions the leadership's prudence, practical judgment, or ability and will to implement desirable goals. The citizen who is concerned with fairness raises the issue of justice. And the citizen who challenges governmental authority raises the issue of legitimacy; one may even ask, "What are the proper functions of my government?"

Wisdom, fairness, authority—when most citizens believe that public policies possess these attributes, the polity enjoys a high degree of popular support. When citizens do not believe that public policies possess these characteristics, dissent erodes consensus. And if protest is expressed in organized group action to replace the political leadership, dissent is transformed into opposition. This is especially likely when questions of justice and legitimacy are raised and when efforts to influence the leadership have failed.

Soviet political leaders have traditionally suppressed dissent and opposition. An armed rebellion by the formerly pro-Bolshevik working-class sailors of Kronstadt was quelled in March 1921. Factions within the CPSU were proscribed at the same time but have persisted. Opposing political parties were eliminated in the early 1920s. The independence of trade unions and other economic, social, and professional organizations was terminated by 1930. Peasants who resisted forced collectivization were slaughtered or starved in the early 1930s. Also annihilated were nationalist guerrillas who opposed Soviet rule in the Baltic states and Western Ukraine in the late 1940s and 1950s. In contrast, the Politburo members who attempted to oust Khrushchev in 1957 were gradually removed from high office and were merely subjected to various forms of official criticism. But in 1966 party leaders responded to growing dissent among artistic, religious, and nationality groups with selective prosecution and increased harassment of leading protesters. The stepped-up repression produced more unauthorized social and political commentary and more expressions of support for the incarcerated dissidents in the burgeoning *samizdat* literature.

During the Brezhnev period the dissident movement spanned a broad political spectrum: deeply religious "Slavophiles" called for a simple rural life, the end of technological borrowing from the West, and the "purification" but not the elimination of authoritarian institutions (e.g., Aleksandr Solzhenitsyn); liberal "Westernizers" advocated the "convergence" of capitalist and socialist systems—pluralist political, economic, and social reforms at home and extensive international cooperaton to alleviate urgent global problems such as hunger, overpopulation, disease, regional wars, excessive defense spending, and nuclear proliferation (e.g.,

Andrei Sakharov); democratic socialists and radical Marxist-Leninists sought to achieve a highly just and productive society within the framework of a less repressive, bureaucratized, and centralized one-party system (e.g., Roy Medvedev); and reactionary "Russites" demanded a return to the militant nationalism and overt anti-Semitism of Stalin and many tsars (e.g., A. Fetisov). Hence, dissident groups and individuals have embraced very different goals, ideas, and hopes, pursuing their goals in different ways and achieving varying degrees of success in bringing about desired changes. Stephen F. Cohen observes that in *samizdat* "an uncensored debate about the political past, present, and future of the Soviet Union is finally taking place," with dissidents also competing "in subterranean ways, inside Soviet officialdom, and even inside the ruling Communist party."[94]

Despite the diversity of the dissident movement, the number of active dissenters has been small. Estimates range from the thousands to tens of thousands, depending on the year in question and on one's definition of "active" dissent. "Passive" dissent is far more widespread, and it manifests itself in private communication and pent-up resentment, as well as in escapist, self-serving, and antisocial behavior (e.g., alcoholism, corruption, and indifference to politics).

Soviet *theory* concerning opposition and dissent has changed little since the time of Lenin. Organized groups within the party and society are still anathema, and dissent is virtually synonymous with disloyalty. Lenin's doctrine of "the leading role of the party" justifies and legitimizes virtually anything CPSU cadres do within the official party rules, and the principle of "democratic centralism" forbids even intraparty criticism of decisions the leadership has made. The 1936 and 1977 Soviet constitutions guarantee to all citizens the freedoms of speech, press, assembly, mass meetings, and street processions and demonstrations. The constitutions are to ensure these freedoms "by the broad dissemination of information and by the opportunity to use the press, television, and radio." But the party-state *gives* such freedoms to individuals, who do not possess or enjoy them as "inalienable" rights. Furthermore, political liberties are to be exercised only on the condition that they further "the interests of the working people" and "strengthen and develop the socialist system" (Article 125, 1936; Article 50, 1977).[95] "In the contrary case," explained Stalin's Commissar of Justice, "the constitution does not guarantee anything to anybody."[96] Khrushchev's chief ideological specialist, in a statement with equally profound implications, declared: "We have complete freedom to struggle for communism. We do not and cannot have freedom to struggle against communism."[97]

It is very difficult for a Soviet citizen to challenge party policies and completely unacceptable to challenge the party's right to engage in certain activities. CPSU leaders have assumed the responsibility to "lead and guide" the country toward communism, have claimed they possess the capacity to do so, and have recruited into the party the men and women they deem best equipped to achieve this goal. Even regional CPSU officials, to say nothing of rank-and-file Communists and non-Communists, feel reluctant to criticize agreed-upon policies. Hence, it is a courageous individual who suggests that a policy should be changed in the light of new circumstances, and an especially courageous, perhaps foolhardy, individual who deems a policy unfair or illegitimate. One who disputes the CPSU leaders' authority to make specific policies assumes a participatory role for the citizenry and posits a relationship between the party and society that constitute fundamental revisions of Soviet leadership theory.

But important changes did take place in relations among the Soviet political elite and, to a lesser extent, between the political elite and the population, beginning in the Brezhnev years. CPSU leaders invited more constructive criticism and more was offered, solicited and unsolicited, from within the party and state bureaucracies, from established economic, military, technical, scientific, and academic institutions and groups, and from the party *aktiv* and all CPSU members. Increased debate on policy issues—some of it public but mostly private—does not necessarily reflect greater "pluralism" or "interest articulation" in the Soviet polity. Instead, it attests to the policy makers' growing need for and desire to accumulate potentially useful information and to consider seriously the recommendations of those who possess special skills, administer national policy, or are directly affected by present policies or policy alternatives under consideration. In a word, the Brezhnev administration cautiously expanded opportunities for *some* citizens to participate more fully in policy formation and implementation and *periodically* to question the wisdom and feasibility of party policies in *selected* issue areas. The collective leadership thus encouraged wider but circumscribed debate about national goals, priorities, and ways to implement public programs more successfully.

What is the significance of these developments for nonparty members who desire fundamental changes in Soviet society? First, top party leaders reasserted the right to distinguish between "responsible" and "irresponsible" criticism and between "loyal" and "subversive" opposition. Second, the Brezhnev leadership specified more precisely the nature of constructive criticism but kept the criteria and standards of censorship flexible and highly politicized. Third, leading CPSU officials regularized procedures for generating policy-relevant information and for exchanging information among national and regional party and state bodies. Fourth, the Politburo, Secretariat, and Central Committee departments accepted and sometimes solicited proposals for incremental change, recommendations concerning the effectiveness and efficiency of specific programs, and information about contextual conditions at home and abroad. Fifth, top CPSU bodies welcomed suggestions only from certain sources, which varied from issue to issue, and only through established procedures and channels, formal and informal. Finally, party leaders and dissidents both found it difficult to distinguish among statements that questioned the wisdom, fairness, and legitimacy of public policies.

Brezhnev increasingly based the legitimacy of the political system on procedural rather than substantive foundations. It became harder and harder to criticize the content of policies because the CPSU leaders had formulated them in a consensual and consultative manner. Also, it became more difficult to criticize policy outcomes without questioning policy-making practices and the authority of the regime. Although the collective leadership vaunted its search for feedback and primary data, it did not find useful or acceptable much of the information elicited from the major bureaucracies and from society. Obtaining problem-related information from lower-level units is difficult in any hierarchical organization, and the tsarist Russian and Stalinist legacy of secrecy in most interpersonal relations further impeded upward and lateral communication. Also, top CPSU leaders debated the implications of constructive criticism, remained chary of complaints about the fairness of policies, and brooked no criticism about the legitimacy of party decisions or decision-making practices.[98]

Herein lies the dilemma for all concerned. On the one hand, Brezhnev and his successors have extended the range of permissible disagreement on many issues and have cautiously attempted to adapt the party's activities to current scientific-technological and socioeconomic conditions. CPSU leaders surely do not want to undermine the party's predominant position in the Soviet polity or to abandon the key principles of "the leading role of the party" and "democratic centralism." *But these very principles, which have long legitimized the Politburo's powers, deny to others the right to oppose, dissent, and advocate significant policy changes*. Any criticism of the general secretary (even by other Politburo members) and any criticism of the "collective leadership" (even by other senior party members) can be labeled "anti-Soviet" or "antiparty." Anyone who expresses nonconformist ideas can be accused of "undermining the authority of the Communist party" or "defaming the theory of Marxism-Leninism." Anyone who violates the unwritten rules of personal and political behavior can be harassed and brought to trial (e.g., individuals and groups trying to exercise certain rights explicitly granted in the constitution). Hence, when Soviet citizens and all but the very highest CPSU leaders strive to change important priorities, policies, programs, and policy-making procedures, they directly challenge deep-seated tsarist and communist political traditions, jeopardize their careers in a society where job alternatives are few and are highly politicized, and even risk incarceration under Soviet law.

Some Soviet dissidents seek to undermine the authority of their party-government, but most want political power used for different purposes. All critics of national policies and policy making weaken the authority of the Soviet leadership because authority is still largely based on a doctrine that claims the CPSU possesses the right to influence virtually every area of human activity. Until the party's authority becomes more solidly grounded on participatory and performance criteria—especially expanded public involvement in decision making, multicandidate elections, a firmer rule of law protecting iconoclasts, a more equitable nationalities policy, improved living standards and social services, scientific and technological advances, and national security and foreign policy successes—Soviet leaders will probably continue to view dissent as a threat to their authority.

How have Soviet dissidents coped with this problem? Beginning in the Brezhnev years, dissidents have responded in at least seven different ways.

The most common response has been to try to criticize only the *wisdom* of party policies. Roy Medvedev declares:

> One of the most serious inadequacies of our present political structure is that it is based on the assumption of total ideological uniformity throughout society. No provision is made for divergent views, and normal dialogue cannot take place within the party or outside. This gives rise to such abnormalities as dismissals, trials, and even psychiatric treatment as a means of political reprisal.[99]

But Medvedev is quick to add that these "defects" can be corrected within the framework of a one-party socialist system and that changes must be made to increase the flexibility, strength, and stability of the *present* polity and economy. Medvedev's argument for freedom of speech is firmly grounded on considerations of prudence and pragmatism:

> The problems which constantly crop up before the leadership of our country—political, foreign, economic, social, technological—are of

> such enormous complexity that there must be a study of alternatives, and different ideas must be tested out even in the political sphere. Yet such experimentation is unthinkable without freedom of debate and discussion, without the clash of different points of view.[100]

In short, Medvedev affirms that "political minorities" are bound to exist in socialist societies and that they should have the right and opportunity to express their views before *and* after decisions are made.

A second common response has been to challenge the *legality* of judicial and secret police actions. Soviet dissidents repeatedly request that investigators, prosecutors, and judges adhere to and implement the letter of *existing* law, in particular the substantive and procedural provisions of the USSR constitution and the criminal codes of the Union Republics. Dissidents argue that the constitution is the highest law of the land and that the *constitutionality* of all actions and other laws should be measured against this standard, as guaranteed in Articles 19 and 20 (1936 Constitution) and Article 74 (1977 Constitution). Also, they note that "Marxist-Leninist theory has not been proclaimed by law as the ideology compulsory for all citizens" and that people who "think differently" or hold unorthodox *convictions* have not committed a crime under Soviet law.[101]

Dissident contentions about the functions of law in a developed socialist society are very important. "A law is a law only if it binds everybody," Valentyn Moroz asserts, implying that the party that *makes* law should not be *above* the law and that governmental agencies administering law (notably the KGB and the procuracy) should never operate outside of the law.[102] But Soviet leaders have traditionally maintained that law is an instrument to serve the state, not to limit it, and that individuals exercise rights conditionally, not unequivocally. In striking contrast, most Soviet dissidents argue that the crucial components of a polity are laws that restrain both the state and the individual, that cannot be changed or flaunted at will, and that are adjudicated by men and women whose authority derives not from the present leaders but from society's commitment to abide by common, usually written, rules. Violations of statutes are therefore considered imprudent, unjust, or illegitimate.

A third response has been to impugn party policy *in light of Soviet ideology*, in particular the ideas and actions of Lenin. The transformation of Lenin's "dictatorship of the proletariat" into the personal dictatorship of Stalin has been severely criticized. In his monumental study of the Stalin era, *Let History Judge*, Roy Medvedev depicts Stalinism as a "serious and prolonged disease" that afflicted a fundamentally productive, just, and legitimate system. Present-day CPSU leaders are exhorted to end all vestiges of Stalinist rule, especially "lawlessness and the abuse of power," which Lenin purportedly would never have condoned.[103]

This argument is significant in connection with the nationalities question. Lenin, chiefly in his later years, contended that Soviet Russia's numerous nationality groups should have an equal right and opportunity to develop their respective cultural traditions. Lenin's ideas contrast sharply with subsequent theory and practice, primarily with Stalin's tenet that the Great Russian people should play "a leading role" in the Soviet multinational state and with Khrushchev's and Brezhnev's selective Russification policies. Dissidents employ a powerful but dangerous weapon when they contend that Stalin and his successors ignored Lenin's explicitly stated prescriptions and policy preferences. To deviate from Leninism is by definition illegitimate in the Soviet Union, and those who suggest that present-day party officials act in this way directly or indirectly challenge the leadership's right to rule.

A fourth response has been to reject *the present-day relevance of Soviet ideology*. The late Andrei Amal'rik dismissed Marxist-Leninist ideology as a self-serving tool of the bureaucratic elites, and he predicted the imminent collapse (possibly through war with China) of the illegitimate, unjust, and ineffectual political system.[104] Sakharov, clinging to his faith in Soviet socialism and in the viability of some one-party systems, nevertheless asserted in 1968: "Any action increasing the division of mankind, any preaching of the incompatibility of world ideologies and nations is madness and a crime."[105] Thereafter, Sakharov more and more questioned the desirability and possibility of creating a world communist system and, in a portentous statement, declared: "The basic aim of the state is the protection and safeguarding of the basic rights of its citizens. The defense of human rights is the loftiest of all aims."[106] In 1974 both Sakharov and Solzhenitsyn openly called for the abandonment of Marxism as the official political ideology of the Soviet Union.[107] Solzhenitsyn was soon forced to emigrate to the United States, and Sakharov was increasingly harassed, indeed hounded, by Soviet authorities.

A fifth and equally significant response has been to deny the *legitimacy of CPSU involvement* in certain kinds of activity. Religious believers, for example, do not acknowledge the party's right to interfere with their public or private worship. The power of the state to close down churches, synagogues, and mosques cannot be challenged, but its authority to suppress religious activity can be and is being opposed, actively and passively. Governmental authority to prevent foreign travel and emigration are two other important disputed areas.

Many creative artists and writers also reject the legitimacy of party interference in their work. Some have appealed for the abolition of all censorship, and Solzhenitsyn declared, "Literature cannot develop in between the categories of 'permitted' and 'not permitted,' 'about this you may write' and 'about this you may not.'"[108] A character in *The First Circle* concludes: "After all, the writer is a leader of the people . . . *a great writer is, so to speak, a second government*. That's why no regime anywhere has ever loved its great writers, only its minor ones."[109]

These terse pronouncements implicitly demand that Soviet institutions give citizens greater freedom of choice and more opportunity to express and develop their creative potential. At issue is one's right to differ or dissent and to follow one's own conscience without fear of reprisal. Indeed, most dissidents believe that the Soviet polity should be more sensitive to "inalienable" human rights, more responsive to divergent views, and more tolerant of "loyal" opposition. Some dissidents even claim that Soviet rule is illegitimate to the extent that it stifles unconventional values and behavior.

A sixth and highly portentous response has been to repudiate the *legitimacy of Soviet hegemony* over the socialist nations of Eastern Europe and of Soviet rule over the non-Russian ethnic groups within the USSR. Both Solzhenitsyn and Sakharov called for the wholesale revision of the territorial borders of the Soviet Union.[110] They argued that Lithuanians, Ukrainians, Armenians, Uzbeks, and other European, Transcaucasian, and Central Asian peoples should have the de facto as well as the constitutional power to secede from the USSR, in order to form separate countries or to be incorporated into neighboring ones. They also contended that dispersed ethnic and religious groups, such as Germans and Jews, should be allowed free departure from and return to the Soviet Union. In short, these dissident views are a thoroughgoing repudiation of the USSR's territorial integrity, with profound implications for its future economic power and international influence.

The seventh and most portentous response has been to condemn both the *theory and practice of Marxism-Leninism throughout Soviet history*. This is precisely what Solzhenitsyn does in his extraordinary *samizdat* work, *Gulag Archipelago*.[111] Detailing the horrors of the Soviet prison camp system from Lenin to Brezhnev, Solzhenitsyn argues that the "excesses" of the Stalin era were *necessary* and *inevitable* consequences of Leninism and of the Marxist ideology that inspired it and on which it was based. Furthermore, Solzhenitsyn insists that the most odious characteristics of the Stalinist period have been preserved virtually intact in the contemporary Soviet polity. Solzhenitsyn expresses unbridled contempt for the past and present Soviet leaders. Most important, Solzhenitsyn categorically rejects the authority of the CPSU, and by his example he encourages others to place loyalty to God and conscience above loyalty to party and ideology, and to act accordingly. Clearly, this is treason from the perspective of the Soviet leadership.

Yet the Brezhnev regime's crackdown on dissent, initiated in the mid-1960s and gradually intensified thereafter, did not signify rejection of all dissident demands. In fact, some petitions, especially those with vocal international support and linked to East-West relations, spurred the party leadership to adjust its treatment of the dissidents themselves (e.g., to allow greater emigration of Jews in the mid-1970s). To be sure, Brezhnev and his colleagues considered many kinds of criticism illegitimate and harmful, especially far-reaching unsolicited proposals that came from unofficial sources in unofficial ways. But the top Soviet leaders expanded the scope of permissible debate among the bureaucratic and educational elites and CPSU members. At the same time, party and KGB officials controlled and suppressed unorthodox political views and behavior and staunchly defended their prerogative to do so.

The dilemma discussed earlier remains. Moroz summarizes the respective predicaments of the Soviet dissident and the party-state:

> The ruling power claims to be the only fount of "the mind, honor, and conscience" of the whole society—and then solemnly proclaims the "politico-moral unity of society." In so far as the Cog [a programmed person] is concerned, the eternal question, "Where to go?," is made into a formula which requires no exertion of the intellect: "Wherever they lead me." A human being deprived of the ability to distinguish between good and evil becomes like a police dog, which is moved to rage only on orders and perceives only the evil that is pointed out to it. . . .
>
> [But] when a despot proclaims his monopoly over reason, honor, and conscience, and forbids anyone to develop these qualities independently, it is the beginning of the spiritual emptying of man. . . . Nothing will replace the free, unregimented thought of an individual whose creative ability is the only motive force of progress. We owe progress to those who have kept their ability to think and have preserved their individuality despite all attempts to erase it. A person without an individuality becomes an automaton who will *execute* every-

> thing but will not *create* anything. He is spiritually impotent—the manure of progress, but not its motor.[112]

Brezhnev's collective leadership perceived aspects of this problem and attempted to cope with them, in part by encouraging certain kinds of creativity, experimentation, and feedback among top executives and specialists, while simultaneously repressing the most influential spokespersons of virtually all significant dissident groups and viewpoints. One can observe two divergent but by no means contradictory trends during the Brezhnev years. On the one hand, the Politburo and Secretariat solicited more and better policy-relevant information and proposals for incremental change from party, state, production, scientific, and educational officials and from rank-and-file communists, thus broadening and deepening elite participation in the formulation and implementation of policies in many fields. On the other hand, the Politburo and Secretariat increasingly stifled unsolicited suggestions and demands for structural or policy changes from unofficial sources and through unofficial channels, thus disheartening and decimating the dissident movement.

The communication system of developed socialism, with its distinctive mix of traditional and modern elements, does not seem to be well suited to the advanced industrial era. But Brezhnev's collective leadership and its successors have expanded the opportunities for judicious and informed criticism of public policies from various sources, and they have done so in a cautious and controlled manner that has not undermined institutional and social stability. Political-administrative continuity has been of the utmost importance to present-day Soviet leaders, and their efforts to preserve the basic features of the polity in an era of remarkable scientific-technological and socioeconomic change have decisively shaped the theory and practice of what we term "technocratic socialism." From a comparative perspective, we will analyze the core components of technocratic socialism in our concluding chapter.

Conclusion: Technocratic Socialism

Throughout this study and in the previous volumes of our trilogy, we have argued that the substance of official Soviet ideology significantly changed and the relationship between theory and practice moderately changed during the post-Stalin period. These developments took place in "grand" as well as "realistic" ideology and among the top CPSU leaders as well as various managerial and educational elites. The present-day Soviet concept of progress includes theoretically interesting and policy-relevant ideas about the interconnections among scientific-technological, socioeconomic, and political-administrative changes. Even when Soviet thinking is highly traditional, it often influences the making and implementation of policies. Social theorists such as Afanas'ev and Fedoseev are high-ranking party leaders who have a considerably greater impact on national politics than do virtually any Western theorists. Also, Soviet conservatives and reformers are debating the nature of the NTR and its consequences at home and abroad, and this ongoing debate, including its ambiguities and lacunae, is highly politicized.[1]

Soviet perspectives on the contemporary era changed more than policies, and policies more than institutions, under Brezhnev's collective leadership. We have examined here and elsewhere the widening disjunctions between the Soviet polity and society and between innovative Soviet orientations and ossified organizational relationships. Our findings underscore the importance of scientific-technological and socioeconomic pressures for change, the CPSU's capacity to sustain remarkable political stability, the improvement of information flows among the top leaders, elites, and masses, and, perhaps most important, the conservatism institutionalized by Stalin's personal dictatorship.

Soviet analysts affirm that the key element of mature socialism is the party-state's use of modern science and technology for traditional purposes.

Top CPSU leaders are to make final decisions on all important questions, control the articulation and aggregation of interests, and set substantive and procedural parameters for public debate. Experts are to participate more actively in policy making because they can help to formulate feasible policies and to legitimize strategic as well as tactical decisions. Significantly altered is Stalin's highly autocratic power over the polity and society as well as Khrushchev's reorganizations of the bureaucracies and mass campaigns in the pursuit of utopian goals. Instead, Brezhnev's Politburo established an oligarchical but consultative policy process that elicited information from the political-administrative and scientific-technological elites in exchange for a considerable amount of career security and a modest amount of cooperation in the implementation of centrally prescribed policies. Brezhnev governed in accordance with selected Leninist, Stalinist, and Khrushchevian values. Basic goals and structures were not open to broader and deeper debate even among the bureaucratic elites. The priorities and operations of the highest CPSU organs were also not subjects for dispute. Questions for public discussion were framed as technical issues concerning the most productive methods of enhancing the common good. In other words, "technocratic counsel" defined and delimited but did not determine or dominate strategic choices.[2]

The theory and practice of societal guidance in the USSR are now primarily *technocratic* in nature—that is, they value technical (instrumental) more than social (symbolic) rationality. Systems of *instrumental* action emphasize technical language and laws that explain and predict social and physical phenomena and that can be empirically reaffirmed, refined, or rejected. The growth and application of analytical knowledge augment political leaders' capacity for technical control. Problem solving consists of selecting the most productive methods of solving centrally defined dilemmas. The aim is to discover means of implementing desired ends and maximizing effectiveness and efficiency through goal-*seeking* feedback. In contrast, systems of *symbolic* action emphasize shared language and widely supported social norms. Equality rather than hierarchy, reciprocity rather than sanctions, and free-flowing rather than constricted information are the bases of such interaction. Norms are grounded on the mutual understandings and interests of a community of people actively engaged in a dialogue on important issues. Ends and means can always be changed, and both are continuously responsive to a wide variety of goal-*setting* feedback.

These two systems of action underscore the differences between *democratic and technocratic rationalization*. Democratic rationalization focuses on negative freedom and goal setting, whereas technocratic rationalization focuses on positive freedom and goal seeking. Democratic rationalization legitimizes expanded public debate about the socioeconomic ends toward which scientific and technological advances are to be directed. Technocratic rationalization legitimizes more and better expert advice about the most feasible means of implementing the national political leadership's goals. Self-selected or elected officials thus enhance their capability to mold society in conformity with *their* interests and with *their* interpretation of the public interest.

Instrumental action can accelerate socioeconomic progress by expanding scientific-technological control over nature and society. But technocratic rationalization does not necessarily lead to greater wisdom, justice, and legitimacy. In fact, increasing technical capabilities can reinforce political leaders' social and psychological domination over citizens. Only symbolic action nurtures social and psychological freedom among all or most elements of the population. Democratic rationalization spurs people's collective and individual search for meaning in their lives and strengthens their polity's and society's capabilities to fulfill these aspirations.

For democratic rationalization to take root, the public realm must emerge as an arena for debate concerning strategic as well as tactical questions. The relationship between the political leadership and citizenry, especially between the politicians and experts, must serve the common good. Through meaningful, multidirectional, and multifaceted communication, traditional values and practices are reassessed and then revised or reinforced. Such dialogue stimulates creativity and initiative whose sources and outcomes are shaped by common values. In turn, values are adjusted by cooperative problem solving, joint exploration of strategic and tactical potentials, and reciprocal influences and mutual learning and respect.

Highly industrialized social orders and their mass polities have been contributors to worldwide trends. Even industrialized democracies have been less concerned with public discussion of the rationality of political actions than with electoral or plebiscitory ratification of governmental decisions. To be sure, modern telecommunications have greatly increased information flows between the leaders and the led. But contemporary science and technology have impeded many kinds of communication between specialists and nonspecialists. Providing the public with policy-relevant information has been hampered by audience manipulation, financial constraints, technical complexity, product obsolescence, information overload, particularistic interests, military secrecy, and other factors. Also, the sciences have become so specialized that virtually all experts know little about developments outside of their own fields. Furthermore, most specialists do not question the foundations of the socioeconomic systems of which they are privileged members. Even nuclear physicists have had little influence on the awesome political uses of their knowledge. Hence, scientific-technological progress does not necessarily lead to socioeconomic progress or does so for the few rather than the many.

An advanced industrial society can be rationalized democratically only if the public is activated. Knowledge must not be accumulated and applied merely to coerce segments of the population with new instrumentalities. According to Jurgen Habermas, a profound danger facing modern societies is the use of information solely or primarily for technical control. When the natural and social sciences

> seek to wrest from contingency that which is empirically uniform, they are positivistically purged of insanity. They control but they do not control insanity; and therefore insanity must remain ungoverned and uncontrolled. Reason would have to rule in both domains; but this way, reason falls between two stools. . . . An exclusively technical civilization . . . is threatened by the splitting of its consciousness, and by the splitting of human beings into two classes—the social engineers and the inmates of closed institutions.[3]

Comparative Perspectives

Although there have been democratic glimmerings in the major Soviet bureaucracies as well as democratic beacons in the dissident movement, the Brezhnev, Andropov, and Chernenko administrations' approaches to rationalization were overwhelmingly instrumental. Symbolic interaction was guardedly expanded to include more bureaucratic elites and "responsible" citizens but not dissidents of any stripe. Soviet theorists claim that scientific and technical rationality fosters progress in the USSR because the means of production are collectively owned and socioeconomic planning is comprehensive. The public interest purportedly flourishes when "progressive" national institutions and programs utilize modern science and technology. The Soviet concept of developed socialism rejects the need for organized groups to question the central guidance mechanisms and to compete with one another in order to *discover* the common good. Hence, instrumental action obstructs symbolic action, and the democratization of Soviet society is forestalled by the party leaders' efforts to expand technical control over the natural and human environments.

The Soviet leadership's focus on technocratic rather than democratic rationalization will now be elaborated, first by comparing Shakhnazarov's forceful presentation of officially sanctioned views with those of the French and Soviet Marxist critics, Roger Garaudy and Roy Medvedev, and then by comparing various Soviet theorists with Western "postindustrial" theorists and political scientists. For a deeper understanding of contemporary Soviet perspectives on politics, society, and technology, one must examine not only what they include but what they exclude and obfuscate as well.

Soviet commentators maintain that technical rationality produces technocracy in capitalist but not in socialist countries and that the NTR fosters socioeconomic "progress" in the USSR by helping a "progressive" polity to satisfy people's material and spiritual needs. This argument is tautological, as the two words quoted above indicate. Also, the Soviet population is viewed as an "object" of improved management rather than as a "subject" of management. Hence, a sharp distinction is drawn between the leaders and the led, with the former periodically allowing the latter—rather than the latter permanently enjoying the right—to debate the nature of progress and the purposes for which technological advances might be used.

Brezhnev's Politburo and its successors have located the necessity for change in the "value-free" NTR and in the "nonpolitical" domain of increased productivity and optimality. Dismissing the possibility of political transformation in the USSR, they insist that the party's dominant role in constructing ideological unity, preserving social stability, and linking scientific-technological and socioeconomic progress is to be maintained. Shakhnazarov declares:

> To renounce [the leading role of the party in the building of socialism and communism] would be tantamount to rejecting the need for political leadership *per se* and would almost certainly lead in practice to anarchy and chaos and eventually degeneration of the system, seriously threatening the accomplishments of the socialist revolution. . . . The actual practice of socialism shows quite clearly that the one-party system is capable of serving as the instrument of social development and ensuring social progress at all stages of construction of the new society.[4]

Nevertheless, Soviet leaders and analysts acknowledge that progress is difficult in the era of the NTR. The more mature the social relations, economy, culture, and way of life of socialist society, the more complicated the task of guiding its development. If the NTR is to be fused with the

Soviet system, the three chief operations of the party—political, organizational, and ideological—must be adjusted and improved.[5]

Soviet theorists have long contended that "scientifically substantiated" policies are needed to manage the problems and opportunities of socioeconomic modernization and that "scientific" policy-making procedures are needed as well. In the Stalinist period, "scientific" policy and policy making were equated with ideological homogeneity and obedience to central directives. Science was synonomous with a clearly articulated body of thought and prescriptions—Marxism-Leninism-Stalinism. In the post-Stalin period, however, ambiguities emerged. The "grand" ideology was still considered to be scientific, but it was increasingly supplemented and supplanted by a "realistic" ideology based on organized knowledge. Shakhnazarov asserts:

> The NTR is introducing more and more entirely new factors into social life, and these are changing existing proportions and concepts, making it necessary to amend perfectly good, well-formed theories as regards methods and schedules for solving various tasks. . . . Although the need for extensive scientific research and for application of its results in order to increase the efficacy of party leadership is present at all stages of socialist development, it manifests itself with special force in developed socialist society. [Officials and theorists] must search for the optimal political structure. . . . The highest, ultimate goal of management, expressed in the most general terms, is optimization of the way the system functions, employing the minimum effort and outlay to achieve the maximum effect.[6]

Hence, the mature Soviet polity is to increase the linkages between theory and practice and the utilization of scientific-technological advances in national policies and policy making. Determining the priorities and assessing the results of research and development remain the prerogative of the party. Organized knowledge sectors are to provide information that helps the CPSU leadership to conceptualize and prescribe feasible policies. Shakhnazarov does not mince words: "Social analysis, extensive application of scientific methods, are the essential bases for *preparing* a correct social policy. But the *actual* policy can only be the concern and responsibility of . . . the Communist Party."[7]

Technocratic rationalization continues Leninist and Stalinist traditions. The pursuit of conscious mastery over recalcitrant historical forces is being adapted to contemporary conditions. Shakhnazarov observes, "As socialist society gradually becomes more mature, there is less and less room for spontaneous, elemental, and uncontrolled processes. This is not because society becomes capable of ignoring objective laws, but because through deeper understanding of them it is able to make them serve its own interests."[8] Also, Stalin's emphasis on ideological unity is being applied to a developed socialist society. Rarely have contemporary Soviet leaders and theorists raised questions such as these: Should organized knowledge sectors help to formulate action projects rather than merely provide the information on which the CPSU leadership acts? Is the NTR generating the need for a fuller synthesis of knowledge and action rather than for a political system sustained by coercion, compartmentalism, complacency, and corruption? Is participation in the making and implementation of decisions, rather than obeisance to the often unclear and contradictory dictates of top party-state bodies, essential to progress? Although a few Soviet officials have gingerly posed similar questions in the post-Stalin period, the issues of symbolic rationality and democratic rationalization were most compellingly expounded by Western and Eastern European Marxists and by Soviet dissidents.

Garaudy emphasizes the "predominance of the most fundamental and specifically human need, the need to create."[9] He calls for a creative upsurge in the capacity of cultures to transform potentials for change into reality and to define and develop needs that become rooted in human thinking and behavior. Marx is for Garaudy a proponent of critical inquiry and an opponent of dogmatic thought. Searching for the source of dogmatic distortions of Marxism, Garaudy locates it in the insufficient autonomy of theory vis-à-vis practice. He contends that ideological thinking often lies in the domain of an already constituted reality or of a categorical rejection of reality, thereby reducing thought to a preference for one kind of preconceived society rather than another.

Garaudy calls for an ongoing cultural revolution. He maintains that the industrial revolution placed culture outside of the realm of work and that industrialized societies foster material rather than spiritual growth. Culture has become a palliative to the frustrations of work, not a motivation to create more productive and satisfying types of work.

According to Garaudy, the cultures of art and science should develop in "dialectical unity."

> I intend to uphold the apparent paradox that aesthetic education is strictly complementary to the great transformation of our time, not a counterpoise to scientific training or an escape from technological civilization, but a major and even dominant factor in education, inso-

> far as the discovery of ends must precede and direct the search for means. . . . We need more than ever an art of invention. The chief virtue to cultivate is not logic but imagination, on pain of reducing culture to the purely operative function of taking the ends for granted.[10]

Creativity thus conceived is the artistic attribution of meaning to phenomena, coupled with the scientific power to explain, predict, and control. Invention is the imagining of alternatives not contained in extrapolations of present trends. By "rupturing" a dogmatically circumscribed present, "prophetic" moments pose fresh alternatives for action and help to "invent the future." Only a society capable of nurturing its innovative capabilities will be able to cope with the complexity, uncertainty, and rapid change generated by modern science and technology.[11] In short, rationalization is viewed as a continuous symbolic challenge, not as an instrumental challenge that has been or can be met by the establishment of a particular socioeconomic or political-administrative system.

Garaudy stresses the interplay between the art of invention and the cybernetization of experience in advanced societies. Monopoly of initiative and decision by national institutions is thought to retard human development because political leaders cannot select just and feasible ends without meaningful participation from the citizenry. Although the burgeoning of technical and normative information is too great for the center to manage, it is to amass, interpret, and synthesize as much information as needed. Determining these needs and applying data to collectively determined social goals are major challenges confronting modern socialist and capitalist states. In order to formulate and implement feasible, effective, and legitimate policies, national leaders are to delegate the power to resolve all but the most essential issues to diverse governmental and public bodies. Management consists "not so much in issuing and enforcing rigid instructions, as in coordinating and orienting a mobile complex of creative centers, all to some extent autonomous and in a state of constant interaction."[12]

Garaudy is not a Western liberal calling for the separation of governmental powers or for the predominance of negative freedom. Instead, he envisions an enlightened centralized polity facilitating dialogue between the center and periphery about strategic and tactical decisions and their implementation. "A feedback of initiative from the bottom replaces the manipulation of men as though they were things devoid of subjectivity."[13] Also, Garaudy favors computerized decision making and underscores the *decentralizing* effects of computer applications. "The use of the computer negates the very principle of hierarchical bureaucracy and substitutes an alternative principle of interdependence."[14] Such interdependence would considerably change institutional power relationships in one-party systems. The center is to encourage and synthesize a broad range of initiatives emanating from society, and the elites and masses are to interact extensively in the formulation of decisions and in the assessment of their consequences.

Soviet analysts attack Garaudy's views, claiming that they advocate the destruction of the party and disregard the laws of dialectical and historical materialism.[15] Garaudy's proposals would allegedly create pluralism or anarchism, making it impossible to guide society in a purposeful and progressive manner. A common ideology, comprehensive and integrated socioeconomic planning, fair-minded aggregation of competing interests, guidance of fragmented state bureaucracies with technocratic tendencies, and mobilization of public support for national goals are deemed crucial to the maturation of a socialist order. The CPSU is to become ever more disciplined so as to direct an increasingly differentiated society developing in a dynamic international environment. Also, individual CPSU organizations are to upgrade their scientific-technological capabilities because of their presumed ability to use the NTR's achievements wisely and the demonstrable need to do so. Hence, the strengthening of the party's "leading" role is justified by the posited benefits of managing centrifugal forces at home and abroad, by the purported rationality of the laws of social development, and by the claim that only one organization comprehends and conforms to these laws and earns the trust of all classes by so doing.

Garaudy retorts that historical laws must be placed in the domain of thought rather than ideology, if the rationality of history is to manifest itself. He emphasizes the need to overcome previous mistakes by creating new laws, which are viewed as hypotheses to be verified, amended, or discarded in practice. Only a socialist polity engaging in a creative dialogue with its society is deemed capable of establishing an open system of developmental options rather than a closed system of developmental necessities. In contrast, Soviet spokespersons consider the laws of social development to be objective or clearly materialized ideals. They criticize Garaudy's understanding of historical laws as subjective. Science is characterized as a unity of truth, with objective trends unfolding in a law-like manner. Natural and social scientists are to strive to understand the specific manifestations of general laws and to uncover the means of implementing laws defining the content and direction of progressive development. The party's task is to govern society on the basis of correct and proven laws; incorrect and unproven hypotheses are to be eliminated from public discussion.

Garaudy's and the official Soviet view of advanced society differ in their understanding of the relationships between freedom and rationality. For Garaudy, the NTR creates the objective conditions under which democratic rationalization can be extended to socioeconomic goals as well as means. For Soviet analysts, the NTR creates the objective conditions under which technocratic rationalization can preserve the Stalinist social and economic legacy. Furthermore, Soviet theory and practice emphasize positive liberty at the expense of negative liberty. Brezhnev and his successors have linked positive liberty with technical rationality, thereby retarding negative liberty and symbolic rationality. Whereas positive liberty can best fulfill the goals of the industrialization and urbanization phase of development, negative liberty can best discover new socioeconomic goals. And, as Soviet reformers contend, the NTR's intellectual, motivational, and institutional complexities increase the importance of negative liberty.

Garaudy underscores the importance of both positive and negative freedom. On the one hand, he affirms that a scientifically directed society satisfies existing cultural, social, and economic needs, establishing the preconditions for the self-actualization of the individual. On the other hand, he contends that the new needs must be continuously conceptualized and fulfilled, expanding the potentials for human development. The political leadership is to integrate the knowledge obtained from specialists with its own values and experience *and* with those of the masses, thereby diminishing the differences between the leaders and the led and creatively linking and developing theory and practice.

For a powerful example of comparable Eastern European thinking, consider the pronouncements by Radovan Richta and his Czech colleagues before the Prague Spring of 1968.

> [The Communist Party must take the lead in evolving] a whole range of new, unorthodox approaches and forms, directed to adjusting technical, economic, sociological-psychological and anthropological conditions for people to engage in *universal creative activity*. Furthermore, existing approaches and forms will have to be rearranged and subordinated to the new. In the spheres of the economy, technology, science, human abilities, etc., power pressures and administrative management assume the quality of external interventions emanating from the superstructure. They served their purpose in fighting the power of capital, large and small, but are incapable of arousing economic activity, stimulating rapid technological advance, and still less of generating scientific discoveries—in

fact, they offer the best way of killing many such prospects. . . .

[Scientific-technical and social] problems can be resolved only when open, free, and friendly discussion takes place. The more the scientific and technological revolution gathers momentum, the more rapidly many accepted ideas will turn into brakes on progress, the greater the intellectual capacity that will be needed to master the new dimensions and modes of movement. Thoroughly elaborated systems of instruments, rules, and methods will be required to facilitate the more rapid and diversified passage through the social organism of progressive findings and ideas that would give the avant-garde elements of social autoregulation [i.e., the Communist party] means of acceleration and would avoid the losses and delays caused by failure to appreciate new methods or by human fallibility.[16]

Roy Medvedev develops some of these themes. He affirms, "A very common prejudice shared by many officials in the government *apparat* and also certain social scientists is that basic economic problems should be solved before attention is paid to the question of democratization."[17] Because of this predisposition, Soviet leaders and their advisors are thought to view technocratic rationalization as a means of avoiding or postponing systemic change. Medvedev declared in 1975:

Committees and commissions of experts are being created, but at the same time undemocratic methods of administration remain intact. In other words, the bureaucrats are being replaced by knowledgeable and more efficient technocrats. This process is developing rapidly in the economic apparat, in the army and security organs, but very much less so in the party or in the press and propaganda departments. Conflict is inevitable between the bureaucracy and technocrats who are far more capable of applying the fruits of scientific progress to the business of administration. Very possibly the major development of the next ten or fifteen years will be the transition from bureaucracy to technocracy. But technocracy, as a distinctive form of "socialist managerialism," cannot resolve the basic problems of Soviet society. There is only one way to deal with them, only one acceptable alternative to bureaucracy, and that is genuine democratization.[18]

Medvedev argues that freer public debate is objectively necessary in the USSR. Freedom is needed to nurture a creative theory–practice cycle and to generate critical thought, not merely technical information, from intellectuals, specialists, and executives. Medvedev maintains that such liberty is blocked by the censors' refusal to distinguish between the development and distortion of Marxist-Leninist ideology. He contends that censorship over all scientific and scholarly publications should cease and that all statements not directly and immediately threatening the survival of the socioeconomic order should be permitted. "My proposals are a call not for peaceful coexistence between opposing ideologies but for struggle between different views and ideas. The principal weapon of this struggle should not be the complete isolation of our people from all non-Marxist opinions and works but, rather, critical discussion of them."[19] Hence, Medvedev favors the continuing primacy of the CPSU for the purpose of fostering symbolic as well as technical rationality. Instead of dominating every major action in a bureaucratic manner, the Central Committee is to be "a general staff concerned with political ideas and methods and with the study and analysis of political moods among the masses; this would enable it to arrive at a party policy in the true sense of the word."[20]

Many Soviet dissidents believe that socialism can meet people's spiritual and psychological needs better than capitalism but that the all-round development of the human personality is possible only in conditions of genuine democracy. Individuals are no longer to be cogs in the command and control system that spawned forced-draft industrialization, collectivization, and social reconstruction in the USSR. With the expansion of secondary and higher education, the citizenry's ability and willingness to make decisions are seen to increase considerably. The scientific-technical and cultural intelligentsias especially are to reflect on the human condition and to debate alternative courses of collective and personal development. In Medvedev's view:

> The Soviet people must have the widest choice, they must have a real opportunity to judge for themselves and make decisions about all questions related to the formation of their mental world. . . .
>
> Majority decisions must be put into effect once they have been adopted. But this should certainly not mean an end to all discussion. Political decisions are not the same as military orders. It often happens that the disadvantages of a certain decision are revealed only in practice. Sometimes it is not until this stage that it becomes obvious whether or not the decision was correct. Therefore, the majority ought to have the right and opportunity to defend its point of view not only before but also after a decision is reached, and this means the right to opposition. The minority cannot refuse to implement a majority decision. But it should not renounce its own views on controversial mat-

> ters. . . . It is crucial to create both within the party and in the country at large some kind of mechanism for normal dialogue between majority and minority, dialogue with dissidents and among the dissidents themselves. . . .
>
> Leadership means more than approving or disapproving suggestions and proposals prepared by the *apparat*. Often it involves reflecting on the political situation and seeking a way out of a difficult position with the leader himself making decisions, making mistakes, and learning from them. Only then can the party judge the capacities of different men and choose the best.[21]

In short, Soviet dissidents affirm that significantly expanded elite and mass participation are necessary to stimulate creativity, to further scientific-technological and socioeconomic progress, and to increase the symbolic rationality of policy making and policies in the USSR.

Official Soviet theories of the NTR invite comparison with Western theories of "the postindustrial society," "convergence," and "technological determinism."[22] Extensively criticizing these Western ideas, Soviet commentators emphasize the organic nature of societies, the primacy of the social relations of production, and the integrated conceptualization of goals and methods (e.g., regarding "developed socialism"). Soviet spokespersons also stress the need for purposeful management of the NTR and its socioeconomic consequences, centralized planning and policy making subject to "public control," a "rational" international and domestic division of labor, and subordination of particularistic interests to nationally determined priorities at different stages of development. Unlike Western theories about the industrialization of pluralist democracies, Soviet theories focus on the problems of "scientizing" the guidance of a complex but unitary society progressing toward the distant goal of "communism."

In contrast to Daniel Bell, Soviet analysts deny the emergence of a "postindustrial" society in mature socialist systems.[23] They argue that a basic continuity exists between the lower and higher stages of industrialization because the Bolsheviks launched an irreversible social revolution in October 1917. NTR theorists affirm that capitalist societies need both a social and production revolution in order to create a more highly industrialized and socially advanced order. Soviet critics dismiss the possibility that Western countries can preserve the socioeconomic features of capitalism while transcending their present stage of industrialization.

In contrast to the "New Left" theorists, Soviet commentators deny the existence of significant developmental problems *common* to highly industrialized capitalist and socialist countries.[24] The New Left has stressed the challenge of eradicating bureaucratization and depersonalization in all nations, but Soviet observers reject the universality of the New Left critique. They insist that the drives toward an advanced stage of modernization under socialism and capitalism are fundamentally different. According to Soviet analysts, the establishment of a socialist order in the USSR, especially the elimination of class conflict and the creation of effective state planning and management structures, is the foundation on which the edifice of a mature socioeconomic system is to be built.

The Soviet approach resembles that of Karl Deutsch, the American political scientist. Deutsch has characterized the advanced stage of development as an information society, not as a postindustrial one.[25] Soviet theorists, in common with Deutsch and in contrast to Bell and the New Left, define the higher stage of industrialization primarily in terms of improving the capacity of societal guidance mechanisms to plan and manage on a national scale and to cope with the extraordinary proliferation of knowledge that affects social consciousness and public policy.

Hence, Bell's distinction between politics, social structure, and culture—and his claim that they "are ruled by contrary axial principles"[26] (legitimacy and equality, efficiency and effectiveness, and self-realization and self-gratification, respectively)—would be uncongenial to virtually all Soviet theorists. Any who might be willing and able to think in Bell's terms would almost surely underscore the innumerable interdependencies and reciprocal relationships among Bell's three realms and would analyze the "nonantagonistic" and "antagonistic" contradictions, not the "disjunctions," in socialist and capitalist societies respectively. A Soviet critic asserts: "Bell's theories are characterized by a substitution of technological relations for production relations; he then proceeds to define the main types of social organizations and institutions in 'postindustrial society' based on technological relations, attributing to them an unsubstantiated independence within society."[27]

Bell contends that both capitalist and socialist nations are becoming postindustrial societies in that they must confront "a common core of problems," especially of a technical and economic nature. But Bell also maintains that different socioeconomic systems respond in various ways to similar circumstances and that the political and cultural outcomes can differ considerably.[28] J. K. Galbraith comes to the much more deterministic conclusion that "convergence between the two ostensibly different planning systems [capitalist and communist] occurs at all fundamental points. . . . The imperatives of organization, technology, and planning operate similarly and, we have seen, to a broadly similar result, on all societies. Given the decision to have modern industry, much of what happens is inevitable and the same."[29]

Soviet theorists strongly object to Galbraith's suggestion that the Western and socialist "techno-structures" are similar and to Bell's implication that the NTR's essence is merely "the centrality of theoretical knowledge." Although Bell distinguishes between the convergence of property relations and of scientific-technical capabilities, most NTR theorists deny that the *perception* or *identification* of major problems (to say nothing of their remedies) could ever be basically similar in socialist and capitalist societies. Soviet spokespersons insist that the relations between the state and the industrial system are fundamentally different under socialism and capitalism.

The trend in contemporary Soviet thinking is toward acknowledging that *some* problems are common to all advanced industrialized societies but that socialist and capitalist systems are responding to these problems in very different ways. Summarizing many basic elements of Soviet thinking, Shakhnazarov observes:

> To be sure, in practice all or nearly all social phenomena bear the imprint of the prevailing system and class ideology. Even such seemingly nonsocial phenomena as technology and technics are no exception. They, too, can [acquire specific characteristics] in distinctive social conditions, giving rise to distinctive social consequences. . . . Still, we must not overlook the fact that [these characteristics derive] not from the intrinsic nature of technology and technics but from the method of applying them in concrete social conditions. The machine and technology are *per se* products of the human brain and human labor. Seen from this angle, they have no social coloring and are inert in the class context. Their spread and the increasing resemblance of production processes is, therefore, no argument in favor of any convergence of the social systems. . . . Past experience has shown that similar problems may be resolved in quite different ways, depending on the social conditions and the aims sought by the prevailing political forces.[30]

Ironically, this interpretation is akin to Bell's view that highly industrialized societies are "converging" along the scientific and technological "axis" but that profound differences may simultaneously develop along the "axis" of property relations and in the social and political "realm." Since the time of Lenin, the justification of Soviet science and technology policy has been founded on such a dichotomization of the USSR's scientific-technological and socioeconomic capabilities. For example, the Brezhnev leadership maintained that the USSR's increased importation of advanced Western technologies was "objectively necessary" in order to narrow the East-West technology gap and to improve the Soviet people's standard of living.

Like Bell, Soviet analysts have relatively little to say about cultural influences on technological innovation. But Soviet observers focus considerably more than Bell on the interaction between ideology and experience and on the "dialectical" relationships between the actual and *desired* social and economic trends of various societies. NTR theorists are keenly interested in the *purposes* technologies serve and would find incomplete, at best, Bell's conclusion that a postindustrial transformation is "the enhancement of *instrumental* powers, powers over nature and powers, even, over people."[31]

Afanas'ev, like Bell and Deutsch, emphasizes the importance of theoretical knowledge in industrial societies. But Afanas'ev and Deutsch assign much greater weight than does Bell to the generation and use of large quantities of integrated scientific, technical, economic, and social information *for the purposes of* increasing the "learning capacity" of the policy-making system, improving the "steering" of society, and determining national socioeconomic goals and policies.[32] Afanas'ev is chiefly concerned with improving the central leaders' capacity and Deutsch with the central, regional, and local leaders' capacity to make informed choices about values, priorities, and programs and to reduce decision overload at the center, to use timely and pertinent data, and to aggregate powerful and conflicting interests. Afanas'ev pays special attention to "conscious" leadership and Deutsch to responsive leadership, and both recognize that societal guidance is taking place under conditions of increasing scientific-technological and socioeconomic complexity, interdependence, uncertainty, risk, and change.

Afanas'ev's thinking has more in common with sophisticated "rational-deductive" views of policy making, such as Yehezkel Dror's,[33] than with the problem-solving strategy of "disjointed incrementalism" that Charles Lindblom has propounded.[34] However, Afanas'ev and other Soviet analysts have gingerly incorporated some of Lindblom's concerns and Dror's "extrarational" factors into their most recent writings (see Chapter 3). This is a clear sign that "mainstream" Soviet theorists are taking a closer look at actual policy making and administrative procedures in the USSR and a less sanguine view of the NTR's manageability and manipulability, especially if traditional bureaucratic practices persist.

Bell, in contrast, focuses on the socioeconomic and cultural contexts in which political decisions are made, and he does not devote much attention to the ways in which political leaders shape their domestic and international scientific-technological environments. Like Bell and Deutsch, Afanas'ev leaves unanswered some crucial questions about the relationships among politics, social structure, and culture, and about how the Soviet political leadership determines its purposes and methods and how they might, must, or should be adjusted during the NTR. Afanas'ev, Bell, and Deutsch affirm the growing importance of theoretical knowledge, but Galbraith, who has much less to say about information, is the most explicit about the acquisition and uses of power. Especially like Bell, Afanas'ev is reluctant to discuss the relationships between knowledge and power, between political and scientific-technical expertise, and between "goal-setting" and "goal-seeking feedback." Hence, Soviet and most Western theorists minimize the political aspects of advanced modernization, in particular the *interests* served by continuity or change in national priorities and institutional relationships.

To be sure, Afanas'ev and Bell insist that politicians, not technocrats, will "lead" or "rule" in advanced industrialized societies and that "the technocratic mind-view necessarily falls before politics."[35] But critical theoretical and practical questions concern the *interrelationships* between politicians and experts, between political and scientific-technological modes of thought and evaluation, and between national policies and bureaucratic interests. For example, which technically trained leaders and which politically astute technicians decide basic questions in developed socialist and capitalist societies? What are the sources of influence of different knowledge elites, and what are the powers of political leaders to resist specialized interests while at the same time utilizing feasible problem-solving approaches and expertise? How pervasive is "technocratic consciousness" among professional politicians and technical specialists, and are there different kinds of technocratic thinking about politics in different societies and segments of societies? In Bell's terminology, might not the "disjunctions among the axial principles" be *narrowed* or resolved in some societies? Could the principle of efficiency in the technical and economic realm begin to dominate the principle of equality in the social and political realm, thereby inverting or merging the ends and means of politics and altering the sources of legitimacy and effectiveness of many polities?

Soviet theorists have not confronted or thoroughly studied such questions, and Bell would perhaps respond that he is chiefly interested in forecasting portentous social, economic, and technologial trends common to all highly industrialized nations. But Soviet analysts might well retort that socioeconomic progress is the only "objective" need of society; that leadership is an increasingly important "subjective" factor affecting progress in the era of the NTR; that party leaders must periodically adjust the purposes and instruments of power; that they have an obligation to use scientific and technological advances to manage political events and trends; that agnosticism on value questions implicitly supports existing ends-means relationships; and that Bell's ethnocentric "postindustrial society" is a none-too-subtle form of ideological rationalization and liberal capitalist propaganda.

Afanas'ev and Bell do not advocate changes in the essential features of their respective political systems, but they underscore the need for major modifications in the secondary features of their polities in order to benefit from and contribute to scientific and technological progress and to cope with its socioeconomic consequences. Each theorist presents a rather traditional and technocratic view of political dialogue and information exchange within and among nations. Afanas'ev and Bell emphasize the centrality of politics in advanced industrialized society, but both deal with the subject of power tangentially or obliquely—for different reasons, no doubt.

The strengths of Bell's theory include its historical perspective on the evolution of ideas, its use of empirically grounded longitudinal studies of socioeconomic developments, its rich description of changes in the social structure and culture of advanced Western nations, and its partial but provocative "agenda of questions" that future societies will probably have to confront. The strengths of Soviet thinking about the NTR lie in its focus on the primacy of politics, on intrasocietal and intersocietal interdependencies, on the importance of information in decision-making and implementation, and on the need to develop a comprehensive vision of the future society *and* of the processes by which it might be achieved. Soviet commentators thus integrate the analysis of ends and means and attempt to conceptualize feasible methods by which "spontaneous" forces can be "consciously" channeled or diverted.

The Core of a Pragmatic Ideology

As we have argued throughout this book, the Soviet approach to advanced modernization rests on several evolving concepts. Probably the three most important are "the scientific-technological revolution," "developed socialism," and "the scientific management of society." Soviet theorists associate universal changes in the productive forces with the value-free NTR, which manifests itself in mature socialist and mature capitalist societies. But global scientific and technological advances are to be combined with the

USSR's distinctive socioeconomic values and institutions. The bifurcation of socialist and capitalist production relations and superstructures makes illegitimate any criticism of and any change in the Soviet polity's fundamental characteristics. According to CPSU spokespersons, the Soviet political system is the first in world history not to serve the interests of a single class. "A state of the whole people" has replaced "a dictatorship of the proletariat," broadening and deepening "socialist democracy." With the improved educational and cultural levels of the citizenry, the growing ideological unity of different social strata, and the reduced disparities between urban and rural life and between intellectual and physical labor, an increasingly participatory *and* harmonious social order is thought to be emerging.

The Soviet theory and practice of developed socialism attempt to adapt Stalinist structures to present-day scientific-technological and socioeconomic conditions. The party-state's domination of society is to be conjoined with increasingly sophisticated methods of political control. To be sure, the NTR generates objective pressures to increase freedom of action in the public arena and to establish regional and local centers of decision making and implementation. But Stalin's successors have delegated authority and encouraged creativity and initiative only to augment the party leadership's capability to carry out its policies more efficaciously.

Although the USSR's developmental strategies have been continuously debated in the Soviet press, major reform proposals from critics of a *current* general secretary have never been tested. Better informed and relatively unfettered discussion of existing policies, not merely participation in their implementation, would probably enhance the Soviet polity's effectiveness and legitimacy. Because of continuing restrictions on public communication about the past, present, and future, the rationality of key Leninist and Stalinist legacies has not been repudiated, and economic growth and productivity—to say nothing of democratization—have been curtailed. Scientific and technological changes have outstripped political and administrative changes, and economic and social changes have fallen in between. While debate about the scientific-technological origins and effects of socioeconomic progress has been encouraged, debate about its political-administrative origins and effects has been stifled. Nonetheless, the scientific-technological, socioeconomic, and political-administrative dimensions of development have become increasingly interdependent in the post-Stalin era, and reformist Soviet officials and analysts have understood this fact especially well.

Soviet commentators acknowledge that scientific management entails many rationalization challenges. Chief among these are the shifts from autocratic to authoritarian centralism, with higher degrees of specialist participation in decision making and implementation, and from command economic to optimal socioeconomic planning, with greater responsiveness to public opinion and market forces. Soviet observers maintain that modern science and technology have a rationalizing effect on the economic and social life of socialist society. The physical environment is to be enriched by new technologies and materials, and negative consequences such as air and water pollution are to be mitigated by centralized planning. Also, scientific and technological innovations are to rationalize politics and administration, and top party-state bodies are to increase their effectiveness and efficiency considerably. Systemic integration, flexible organizational linkages, and improved technical and normative communication are thought to be key elements of rationalization. Pertinent and timely information flowing to all decision-making points, together with centralized determination of the policy implications of scientific-technological and socioeconomic developments, are viewed as crucial means of enhancing the polity's capabilities.

The Soviet conceptualization of mature socialism emphasizes such technocratic forms of rationalization rather than democratic rationalization. The latter would significantly expand public debate about the wisdom, fairness, and legitimacy of national policies and policy making and would produce a much broader consensus about ends and means. Democratization rests on the equalization of opportunities to advance competing interests, the dispersion of power, and the selection of political leaders who will judiciously choose among different collective and individual developmental options. Soviet claims about socialist democracy notwithstanding, official perspectives are a distinctly authoritarian and technocratic response to the challenge of modernizing the political system of a large industrialized nation.

The conjunction of technical rationality with historical irrationality is the essential meaning of technocracy in an advanced society. Western analysts often view technocracy as rule by technocrats and as the domination of expertise in the political arena.[36] But leadership in a complex and dynamic society molds diverse factors into policy, continuously reassessing priorities and monitoring and adjusting policy outcomes. Developed societies are characterized by an intricate and shifting division of labor with a concomitant fragmentation and evolution of expertise. There are no experts who can make systems-oriented decisions, especially the authoritative allocation of values and the aggregation of powerful competing interests. There is no expertise, emanating from scientific-technological, ideological, or religious sources, on which to base comprehensive, synthetic, and legitimate decisions. Such decisions cannot be formulated on the basis of analytically unconnected knowledge, nor can they be indefinitely avoided.

In distinguishing among modern societies, the basic question is not "*Who* will rule?" Rather, one must focus on the *goals* of different groups in society, on whose *interests* are being served, on how scientific and technological advances are being generated and used, and on which scientific and technological advances are central to political-administrative and socioeconomic development. Because the NTR increases the need for planning and management on a national scale, either generalists or technocrats *acting* as generalists will govern. "Dual executives" at the CPSU's national, regional, and local levels selectively utilize science and technology to integrate and direct society, preserving traditional values and interests in rapidly changing domestic and international conditions. But the same scientific discoveries and technological innovations can be used for different purposes and can have different effects in various segments of the Soviet polity, economy, and society.

Technocracy, then, does *not* refer to the "requirements" or "imperatives" of scientific and technological progress. Instead, technocracy consists of *reciprocal influences* between political-administrative goals and scientific-technological achievements. Particularly important is "reverse adaptation," the adjustment of collective and individual goals to the scientific-technological means available. The criteria and standards of technical rationality become imbued in institutional thinking and action. Also, organizational performance is evaluated in terms of procedures and instruments rather than by the fulfillment of societal needs. Under such circumstances, Langdon Winner observes, "efficiency takes on a more general value and becomes a universal maxim for all intelligent conduct. . . . *Beyond a certain level of technological development, the rule of freely articulated, strongly asserted purposes is a luxury that can no longer be permitted.*"[37]

A sizable gap between ambitious ends and limited means produced Khrushchev's "hare-brained scheming." But a close correlation between ends and means produced Brezhnev's technocratic rationalization. Brezhnev's Politburo strove to preserve the power of the bureaucratic elites and to enhance their control over the social and physical environments. The collective leadership's preferred instrumentalities were modified methods of planning, decision making, and management, especially regularized information flows, systems approaches, and computerization. Although the distinctions between ends and means were small, the available and anticipated means significantly influenced, and in some cases *became*, the ends.

"Technocratic socialism" is the by-product of the technocratization of a socialist system. Seven core components of this concept will now be outlined. They have emerged from rather than guided the analysis and evidence in our book. Furthermore, they constitute the most intellectually influential and policy-relevant Soviet ideas about politics, society, and technology in the post-Stalin period. Leading CPSU officials and theorists have articulated most of these perspectives. Party spokespersons (some more than others) might challenge our interpretation of the nature and significance of the remainder. And Soviet commentators will no doubt reject our overall characterization of their views as elements of "technocratic socialism." Nonetheless, the argumentation and documentation in the preceding chapters, as well as in the other volumes of our trilogy, undergird the following synthesis of the Brezhnev, Andropov, and Chernenko leaderships' worldviews.

First, *science and technology are crucially important means of achieving social and economic progress*. The capability of a political-administrative system to use scientific-technological breakthroughs in the pursuit of progressive aims attests to the growing rationality of that society. Rationalization is rooted in both the subjective commitment and the objective need to improve the guidance of an advanced social order. The authority of political institutions over collective and individual behavior is to be strengthened. Mature socialism adapts socioeconomic planning and management—not socioeconomic values—to the opportunities and problems generated by a global NTR. Institutional changes, if any, are to be incremental.

Genetic and social engineering dramatically increase the power of political leaders for good or ill. Such efforts to mold hereditary and learned behavior constitute progress in socialist societies because their directive capabilities considerably expand by bringing the biological and social sciences into closer accord.[38] Negative spiritual and material effects of the leadership's manipulation of the citizenry are dismissed or deemphasized. Scientific and technological advances are to control the mainsprings of human nature, rather than to enhance people's motivation and ability to choose among new and traditional values. Only in the communist society of the distant future will "creative labor predominate and combine intellectual, emotional, and physical activity."[39]

Second, *political values are to have a mounting impact on research and development and on the physical environment*. With the technification of science and of socioeconomic systems, natural forces are to be manipulated in the light of political decisions, and the scope of politics is to widen.[40] Soviet analysts stress that the CPSU should increasingly stimulate scientific-technological advances and direct the actual uses of discoveries and innovations. Gerasimov observes, "In the contemporary world, man's influence on the natural environment is justifiably compared to the action of crucial geological factors. . . . The NTR holds out to mankind in the coming historical period unprecedented technical opportunities for cardinal changes in nature on a regional and global scale. These changes, however, may be directed toward rational utilization and improvement of the natural environment for the benefit of man or bring disaster to mankind."[41]

Soviet observers contend that the critical variable affecting progress is scientific management. According to L. P. Lysenko, "Managing the interaction of science and society requires the creation of a complex managerial system. Society must 'survey' the changes in all of the components in the social-technical-natural spheres, in order to enhance its capability to manage them. Under socialist conditions, science must transform nature into an object of scientifically substantiated economic activity."[42] For example, the industrialization of outer space is deemed a great opportunity and the militarization of space a great danger. Because humankind's survival depends on managing the competition in space, the rational development and utilization of nonterrestrial resources is to be grounded on international scientific and technological cooperation.[43]

Third, *the integration of the polity and society is the main course of evolutionary advance*. Political institutions are to increase the dynamism and harmony of the existing socioeconomic system. Traditional problems are to be resolved by the NTR, and problems created by the NTR are to be resolved by innovative research and development that fulfills national goals. According to S. I. Popov, "the scientific guidance of social development and the broad application of scientific achievements *in all spheres* of production, culture, and everyday life comprise an objective law of the functioning of developed socialist society."[44] And, as Fedoseev declared in 1977, "A major task of mankind today is to learn how to tackle on a planned and purposeful basis the new economic and social problems *constantly* posed by the NTR."[45]

The scientization of the polity and society are the twin components of the drive for systemic unity and managerial synchronization. G. N. Volkov conceptualizes the future communist order as "a society with the completely automated production of an abundance of material benefits. Such production will comprise a Single Automated System, in which all industries together with agriculture are united and centrally managed according to a single plan that ensures the greatest efficiency. This will be a united, socially homogeneous society because distinctions between mental and physical labor, between town and country, will have been erased."[46] Under present-day developed socialism, in Afanas'ev's view, scientific management must "ensure the integrity of the system, assimilate or neutralize internal and external disturbances, and continuously improve the structural and functional unity of the system."[47] More specifically, scientization includes the improved professional qualifications of decision makers, the increasingly sophisticated organizational technology at their disposal, and the use of systems approaches in management.

Fourth, *top policy-making bodies are to improve their capability to synthesize information relevant to political action*. Although the party's "leading role" is to be strengthened, the NTR creates pressures for a more consultative relationship between political-administrative and scientific-technological elites. The tension between the centralization of power and the decentralization of policy-relevant information is a major "nonantagonistic contradiction" to be resolved.

Soviet theorists argue that it is essential to comprehend and influence the unfolding of objective trends in concrete circumstances. To do so, the party leadership is to inject increasing amounts of timely and reliable information into all stages and levels of decision making. The gathering, processing, and use of information are to be improved by direct contacts with the masses through party committees and public organizations; new methods and technologies such as survey research and computers; and counsel and data from experts such as philosophers, sociologists, economists, statisticians, mathematicians, jurists, psychologists, scientists, engineers, agronomists, and industrial and agricultural production executives. Afanas'ev has long maintained, "A scientific system of directing society can be evolved and successfully applied in specific spheres of social life, and concrete forms, ways, and means of administration conforming to present-day requirements can be worked out, only through the joint efforts of specialists in the most diverse fields."[48]

The CPSU is to solicit and assess expertise in order to clarify and resolve the most important problems facing society. Optimal policies are to be formulated on the basis of scientific-technological and socioeconomic data generated by specialists and interpreted by party leaders in conjunction with political values and circumstances. Afanas'ev reiterates, "Criticism is not an end in itself but a means of perfecting the entire system of management. Criticism can be effective only when it has a constructive purpose."[49] The CPSU leadership is to determine such purposes, thereby ensuring the successful functioning and development of the existing policy-making process, which cannot be sustained and improved without abundant information about current events and trends in all spheres of social life at home and many abroad.

Fifth, *the policy-making bodies are to formulate a developmental strategy and to supervise its implementation.* Social and economic planning and forecasting are to become more comprehensive and integrated and are to incorporate and anticipate research and development priorities. Socioeconomic and scientific-technological trends unifying society are to be fostered; trends disaggregating society are to be reversed. Old and new nonantagonistic contradictions are to be ameliorated, and decision-making organs are to discover and control the evolution of society as a whole. According to Afanas'ev, "It is not easy to find basic links. To do so, one must have perfect knowledge of the entire range of tasks confronting the system and the tendencies of its development and have mastered the art of seeing into the future. To manage is to forecast—to know ahead of time what the course of events will be, in what direction the system will be developing, and what tasks, though not very important at the moment, will acquire decisive significance and play a key role."[50] Hence, the CPSU leadership is to induce the polity and society to be more responsive to one another while stimulating and applying major scientific and technological advances. "The organs of management must be able to sense changes in the system and influence it accordingly, ensuring not only its dynamic equilibrium but its continuous improvement and development as well."[51]

Sixth, *"reverse adaptation" characterizes the relationship between scientific-technological means and socioeconomic ends.* Collective and individual needs are defined with reference to the techniques and instruments available to planning organs, and the creation and especially the utilization of technologies are deemed a primary social need. Because systemic integration is vital to the fulfillment of such needs, the construction of a scientized guidance mechanism is the raison d'être of a developed socialist society. Hence, CPSU spokespersons equate mature socialism more with the *procedures* of scientific management than with the *substance* of traditional Marxist values.

In present-day Soviet politics the blurring of ends and means limits the applications of scientific and technical knowledge and fosters a bureaucratically dominated synthesis. Research and development conforming to a bureaucratic mentality eschew controversial issues and innovative approaches as well as a commitment to the common good. When scientific and technological advances serve the particularistic interests of organizations or regions, the national interest invariably suffers. If Soviet departmental or local pursuits shape the nature and uses of expertise, the growth of knowledge is much less likely to liberate humankind than to repress or alienate it. Indeed, a technocratic regime can exert considerable power over its population. According to Sakharov, modern technology and mass psychology provide more and more opportunities to manipulate the convictions, strivings, and behavior of people. He concludes: "We must be clearly aware of the awesome danger to basic human values and to the meaning of life that may be concealed in the misuse of technical and biochemical methods and the methods of mass psychology. Man must not be turned into a chicken or a rat, as in the well-known experiments in which elation is induced electronically through electrodes inserted into the brain."[52]

Seventh, *elite participation in policy making and mass participation in policy implementation are to be enhanced*. The Soviet citizenry is continuously exhorted to understand and comply with the leadership's interpretation of the national interest rather than to help formulate or alter this interpretation. To be sure, there was a nationwide discussion of the proposed Soviet constitution in 1977, followed by some changes in the final document. Also, letters from the public have long flooded the offices of major newspapers and party committees, prompting increased and regularized replies beginning in the Brezhnev years. Moreover, the responsibilities of the standing committees of the national and regional soviets increased considerably under Brezhnev, giving deputies greater opportunities to contribute to the legislative process in many issue areas. But such participatory activities almost always concerned the politics of details, not the politics of principles. When party officials and citizens communicated through the media, they rarely implied that even incremental changes were needed in the polity's fundamental characteristics. Recommendations for more substantial changes were sometimes disseminated through the specialized press. However, the legitimacy of all complaints and the feasibility of all proposals were determined by the highest CPSU organs. Because of the known and anticipated interconnections among political-administrative, socioeconomic, and scientific-technological changes, and because of the powerful interests vested in the bureaucracies, Stalin's successors have chosen to err on the side of conservatism in expanding the parameters for critical discussion.

Furthermore, the Soviet public still has no voice in choosing or replacing the self-selected party elite to whom it must petition. Single-candidate elections are for government offices only. As a former Soviet lawyer affirms,

> The present Soviet constitution states that the power belongs to the people, who exercise it through elected soviets—councils—and the government of the country is carried out by the Council of Ministers and other administrative bodies. The fact is, however, that true power in the Soviet Union belongs to the apparat of the Communist Party, and it is the members of that apparat who are the true leaders of the country. . . . [Their] power encompasses all spheres of public and private life; it is just as absolute on the national level as it is within each district, each region, and each Union Republic. . . . [It is] a power unbridled by the principle of subordination to the law or by a free press or by the voice of public opinion.[53]

The citizenry's enhanced control over political institutions and personnel would spur democratization in the USSR, but society lacks mechanisms to ensure the responsiveness of the polity. Although Soviet analysts presume that the CPSU's actions are synonomous with the national interest, public opinion research belies the party leaders' claim that their policies and policy-making procedures are enthusiastically supported by all segments of the population. To be sure, the use of more expertise in the policy process has heightened interaction between the Soviet political system and the scientific-technical intelligentsia. But greater social control over the party-state is the obverse of greater party-state control over society. The former constitutes democratization, the latter technocratization. In the former there is considerably expanded elite and mass participation in the formulation of shared goals; in the latter there is considerably expanded elite and mass participation in the implementation of centrally prescribed goals.[54]

Technocratic rationalization has been the hallmark of Soviet politics after Stalin, but democratic rationalization has not always been repressed. An ongoing debate between proponents of the traditional industrialization and advanced modernization strategies has been conducted through the mass media and private party channels for more than three decades. Although the debate has been largely technocratic, some of its participants have cautiously questioned the wisdom of preserving basic characteristics of the Soviet polity, such as the concentration of economic decision-making power at the CPSU's uppermost echelons. Also, the Brezhnev, Andropov, and Chernenko leaderships consistently encouraged judicious discussion of ends-means relationships and voiced diverse opinions about the most feasible methods of achieving traditional goals. The chief result of these efforts has been "reverse adaptation," but glimmerings of procedural and substantive democratization have appeared as well.

Ideology and Politics

Having described and analyzed Soviet perspectives on the advanced industrial era, we offer some concluding remarks about the origins and effects of these perspectives.[55] Above all, modern Soviet ideology has influenced the behavior of the political elites. It has performed both legitimizing and self-legitimizing functions as well as serving as a "guide to action." In addition to CPSU officials' ritualistic affirmations of dogma, they are striving to elucidate and resolve dilemmas and are advocating alternative ways of modernizing the Soviet system under rapidly changing domestic and international conditions. Contemporary Marxist-Leninist theory sets the perceptual and normative parameters and provides the language for pursuing such ends. Also, it incorporates values such as social and institutional stability—even job security for bureaucrats—that were not included or stressed in traditional Marxism-Leninism. Because official Soviet ideology is becoming increasingly pragmatic, it may be influencing considerably—and it has the potential to influence greatly—the party leadership's definition and assessment of political realities and policy responses to them.

If one views ideology as a cluster of utopian goals, then ideology under Brezhnev was much less important than under Khrushchev. But if, as we do, one views ideology as a cognitive approach to the identification, analysis, and resolution of complex public policy issues, then ideology under Brezhnev was even more important than under Khrushchev. Although Khrushchev's utopian ideas influenced *his* actions, Soviet conservatives and modernizers opposed the general secretary on pragmatic grounds. Since Khrushchev's ouster, debates among the bureaucratic elites have been conducted on a pragmatic plane, not on utopian or noninstrumental planes, and programmatic statements have been less grandiose but more policy-relevant. According to Chernenko, the forthcoming third CPSU program "should give a realistic and comprehensively balanced description of developed socialism, [proceeding] from what has been proven in practice, [emphasizing] goals that can be attained by the generations living today, [and reflecting] not only the life that has been changed by us but also our changed political language—a businesslike language that is strict in its phraseology."[56]

In other words, the Brezhnev, Andropov, and Chernenko leaderships were in fundamental agreement about the functions of ideology, especially the need to focus on ends-means relationships, and about the mode of ideological debate, especially conceptual clarity. But the content of Soviet ideology remains very much in dispute, and theorizing continues to be a part of highly politicized efforts to preserve, reform, or discard primary or secondary features of the Soviet system. For example, the possibility of *antagonistic* contradictions under developed socialism was vigorously debated during the extensive national and regional CPSU personnel changes in the early and mid-1980s. The officials and theorists who warned against mounting contradictions favored institutional reform and their opponents resisted it. And recent unrest in Poland was interpreted quite differently.

Although top Soviet leaders are the chief beneficiaries of their system, they also want to change selected aspects of it, especially the performance of the economy and the insularity of the myriad bureaucratic and production units. The Brezhnev, Andropov, and Chernenko approaches to socio-economic modernization constituted low-risk initiatives to integrate a "cellular" society.[57] But poor technical communication, especially lateral communication, has long enhanced the powers of national CPSU officials while reducing the capabilities of society as a whole. Until trust and cooperation among organizations and groups are somehow increased, departmentalism and localism will thrive. And until a Politburo risks the dispersion of political power in the pursuit of economic growth and productivity, "systems approaches" and "science-technology-production cycles" will exist mainly in the minds' eyes of theorists and would-be "scientific" managers.

Technocratic socialism is chiefly an orientation of CPSU and specialist elites; it is not an integral part of bureaucratic behavior or mass political culture. This maturing elitist outlook recognizes that modern science and technology are enormously relevant to domestic and foreign policies. It rejects Khrushchev's optimistic view that the USSR's scientific and technological capabilities will quickly surpass the West's. Instead, it prompted the Brezhnev administration's decision to have many Soviet scientists and engineers copy or adapt Western technologies in order to prevent East-West technology gaps from widening. Acknowledging the difficulties of spurring indigenous technological innovations and creative applications of native and imported technologies, Brezhnev's collective leadership tried to reduce the Soviet citizenry's "highly developed aversion to risk" and "very heavy reliance on personal ties and relations for communications and the conduct of business." But all of Stalin's successors have failed to reduce the fragmentation of institutions and the differentiation of social strata. The probable reason is that the Soviet leadership shares with society "an intense preoccupation with economic and political security."[57]

Everything considered, theories of the NTR and NUO have played an increasingly conservative role since the early 1970s. But, as Stephen F. Cohen observes, "Marxism-Leninism is an unreliable conservative vehicle because it is an ideology, even in its dogmatic version, based upon the very idea, desirability, and inexorability of change."[58] And, as Alvin Gouldner affirms, ideologies "are not simply order-maintaining, *status quo*–enforcing symbol systems [that] reduce certain tensions and provide an *apologia* for certain interests, but, also, and everywhere, undermine 'what is' because they provide a ground for its *critique*." This is especially true "in all socialist countries" because of an unresolved "*basic contradiction: their culture is egalitarian but their social structure is hierarchical; their ideology calls for workers' control of the forces of production, but these are actually controlled by the party and the state*."[59] Hence, contemporary scientific-technological advances, coupled with traditional Marxist-Leninist doctrine, have laid the ideological foundations for major policy and policy-making changes, if and when future Soviet leaders should choose or be impelled to make them.

To date, Soviet theorists have sidestepped such issues and have stressed the importance of centrally controlled change. In the long run, "It is the aim of communism to achieve the highest degree of social organization, the highest social antientropy. It is the object of man's scientific and technological activity to develop ways and means for the antientropic remaking of nature and society."[60] In the short run, however, both the managerial inefficiencies of traditional state socialism and the fissiparous tendencies of developed socialism are to be overcome by technical rationalization. Briefly stated, Brezhnev and his successors have viewed technocratization as the most feasible alternative to the "subjectivism" of the Stalinist and Khrushchevian legacies and to the "anarchism" of dissidents propounding various types and degrees of democratization.

The Brezhnev leadership expanded public discussion about the NTR and NUO in an effort to clarify the *meaning* of "scientific" politics. Brezhnev and his colleagues were well aware of the ambiguities inherent in the core concept of "science" and of the tensions generated by the juxtaposition of Stalinist and post-Stalinist interpretations. But, despite significant changes in Soviet views of scientific-technological, socioeconomic, and political-administrative potentials, the fundamental values of the bureaucratic elites shifted very little. New technologies and problem-solving approaches did not substantially alter traditional socioeconomic and political-administrative structures and behavior. As Gouldner observes, "Scientific and technological expertise rationalize and legitimate only the instrumental *means* used to achieve the organizational goals given, but *not the goals* themselves. These can only be legitimated by value systems and ideologies to which the controlling administrators may link their organizational directives. . . . Even within the most modern bureaucratized state apparatus, science and technology thus operate within limits set by *ideology and interest*."[61] And Kendall Bailes, assessing the Lenin and Stalin periods, affirms: "Similarities in technology adopted by industrial societies do not necessarily determine the kind of social relations that may emerge. Quite the contrary. Differing cultural traditions, ideological climate, and social structure at the time of adoption are crucial factors."[62]

Likewise, Soviet political leaders and social theorists in the post-Stalin period have rejected technological determinism. They have concurred with the Western view that "the challenge to research and inquiry—and eventually to policy—is to find out why the old economic and political forms are not working and to modify them or design new ones that will both preserve the fundamental values of society and yet be adequate to modern technological realities."[63] Worldwide scientific, technological, economic, and social developments have altered the perspectives of top CPSU officials but not their traditional goals and behavior. Portentous changes in the content and functions of ideology as well as cautious adjustments in policy making and administration have accompanied the collective leaderships' efforts to comprehend and cope with the NTR. Neither a production nor a social revolution has taken place. Politicized orientations and choices, not the inexorable demands of scientific-technological and socioeconomic forces, have decisively influenced Soviet domestic and foreign policies. Technological "spontaneity" has shaped but not mastered political "consciousness."

Notes

Introduction: A Soviet Ideology of Advanced Modernization

1. Cyril Black, "New Soviet Thinking," *The New York Times*, November 24, 1978, A27.
2. D. M. Gvishiani, "The Scientific and Technological Revolution and Progress," in *Historical Materialism: Theory, Methodology, Problems* (Moscow: Social Sciences Today, 1977), 41 (emphasis added).
3. B. M. Kedrov, "Criteria of the Scientific Revolution," in *Soviet Studies in the History of Science* (Moscow: Social Sciences Today, 1977), 51, 67, 70–71. On Kedrov's remarkable career, see Werner Hahn, *Postwar Soviet Politics: The Fall of Zhdanov and the Defeat of Moderation, 1946–53* (Ithaca, N.Y.: Cornell University Press, 1982), 161–85.
4. L. I. Brezhnev, "Central Committee Report," *Pravda*, February 24, 1981; translation, *Current Soviet Policies VIII* (Columbus, Ohio: Current Digest of the Soviet Press, 1981), 21 ff.
5. See, e.g., *Chelovek-nauka-tekhnika* (Moscow: Politizdat, 1973); translation, *Man, Science, Technology: A Marxist Analysis of the Scientific and Technological Revolution* (Moscow-Prague: Academia, 1973); *Nauchno-tekhnicheskaia revoliutsiia i obshchestvo* (Moscow: Mysl', 1973); *Nauchno-tekhnicheskaia revoliutsiia i sotsializm* (Moscow: Politizdat, 1973); *Partiia i sovremennaia nauchno-tekhnicheskaia revoliutsiia v SSSR* (Moscow: Politizdat, 1974); and S. V. Shukhardin and V. I. Gukov, eds., *Nauchno-tekhnicheskaia revoliutsiia* (Moscow: Nauka, 1976).

   Unfortunately, few Soviet studies about the NTR and NUO have been translated into English, and many of the translations contain intermittent deletions, additions, or free rendering of the original Russian. Most translations in the present book are our own. When English texts exist we have double-checked the original and have made necessary corrections. Also, we have cited all English-language translations of which we are aware.
6. *The Fundamentals of Marxist-Leninist Philosophy* (Moscow: Progress, 1974), 9.
7. Erik P. Hoffmann and Robbin F. Laird, *The Politics of Economic Modernization in the Soviet Union* (Ithaca, N.Y.: Cornell University Press, 1982).
8. Erik P. Hoffmann and Robbin F. Laird, *"The Scientific-Technological Revolution" and Soviet Foreign Policy* (Elmsford, N.Y.: Pergamon Press, 1982).
9. We offer here a book-length study of issues raised on pages 59–73 of our *The Politics of*

Economic Modernization in the Soviet Union and on pages 7–14 of our *"The Scientific-Technological Revolution" and Soviet Foreign Policy*. We deemphasize economic reform, foreign policy, and military affairs, which we have discussed at length in the first two volumes of this trilogy, in a recent book, and in two comprehensive anthologies. See Robbin F. Laird and Dale Herspring, *The Soviet Union and Strategic Arms* (Boulder, Colo: Westview Press, 1984); Erik P. Hoffmann and Robbin F. Laird, eds., *The Soviet Polity in the Modern Era* (Hawthorne, N.Y.: Aldine Publishing Co., 1984); and Erik P. Hoffmann and Frederic J. Fleron, Jr., eds., *The Conduct of Soviet Foreign Policy*, 2d ed. (Hawthorne, N.Y.: Aldine Publishing Co., 1980). Also, we deemphasize Soviet economic planning, which is the subject of an extensive Western literature, and Soviet science, which is discussed in Loren Graham, *Science and Philosophy in the Soviet Union* (New York: Vintage, 1974); and Linda Lubrano and Susan Solomon, eds., *The Social Context of Soviet Science* (Boulder, Colo.: Westview Press, 1980). Furthermore, we have devoted little attention to technological developments under Lenin, Stalin, and Khrushchev because the interwar years are the focus of Kendall Bailes's study, *Technology and Society under Lenin and Stalin: Origins of the Soviet Technical Intelligentsia, 1917–1941* (Princeton: Princeton University Press, 1978), and because key issues from 1928 to 1975 are examined in Bruce Parrott, *Politics and Technology in the Soviet Union* (Cambridge, Mass.: MIT Press, 1983). Finally, we have not scrutinized the innovation process in specific industries because Ronald Amann, Julian Cooper, et al. have done so in *Industrial Innovation in the Soviet Union* (New Haven, Conn.: Yale University Press, 1982), as has Joseph Berliner in *The Innovation Decision in Soviet Industry* (Cambridge, Mass.: MIT Press, 1976).

For comprehensive studies of the post-Stalin period, see Seweryn Bialer, *Stalin's Successors: Leadership, Stability, and Change in the Soviet Union* (New York: Cambridge University Press, 1980); Jerry Hough and Merle Fainsod, *How the Soviet Union Is Governed* (Cambridge, Mass.: Harvard University Press, 1979), 192–576; and George Breslauer, *Khrushchev and Brezhnev as Leaders: Building Authority in Soviet Politics* (Boston: Allen and Unwin, 1982).

10. Fundamental and stable elements of Soviet Marxism-Leninism include its conceptualizations of "politics," "political system," and "political analysis." Authoritative definitions are:

> *Politics*—the area of activity covering relations between nations, classes, and other social groups and directed at winning, retaining, and exercising state power. The content of politics is always ultimately determined by the interests of a class or an alliance of classes. Any social problem acquires a political character if its solution is related, directly or indirectly, with class interests, with the problem of power (see K. Marx and F. Engels, *Collected Works*, Vol. 3, Moscow, 1975, 120). The most essential element in politics, Lenin emphasized, is the organization of state power; politics is participation in affairs of state, the determining of the forms, tasks, and substance of its activity (see V. I. Lenin, *Collected Works*, Moscow, Vol. 19, 121–22).

From "Our Glossary," *Social Sciences* (Moscow) (hereinafter *SS*), 2 (1978), 309.

> In contrast to the bourgeois interpretation of the theory of systems, the Marxist dialectical approach goes further than separating a given system from the

environment and studying its interconnections with the environment and within itself, and further than identifying the basic variables and determining the goals, alternative forms, and mechanism of activity. It involves in addition, first, looking at functioning as an element of the system's development; second, taking into consideration the distinctive characteristics of a social system as compared with a biological or cybernetic system, on the one hand, and a political system as a special case of the social system, on the other; third, the study of conflict from the standpoint of the unity and struggle of contradictions and opposites; and, fourth, seeking out the main factors of change in the political system which stem primarily from the economy and the structure of society.

Hence, Marxists understand a political system to be a relatively closed system that ensures integration of all the elements of a society and the society's very existence as a single organism, whose heart is the state, which expresses the interests of the economically dominant classes. A political system consists, above all, of political institutions (the state, law, parties, public organizations, etc.), and it also includes a system of communications linking the members of the society, its classes, and other social groups with the political power.

From F. M. Burlatskii, "Political System and Political Consciousness," in *Time, Space, and Politics* (Moscow: Social Sciences Today, 1977), 53–54.

11. Our study is grounded on W. I. Thomas's classic assumption that "situations defined as real are real in their consequences"; Robert C. Tucker's contention that "the political process is influenced by many a material factor, but it has its prime locus in the mind"; and T. H. Rigby and R. F. Miller's observation that "the strength of ideological modes of thinking among Soviet leaders, at least concerning elements of official doctrine, is usually insufficiently appreciated in current Western assessments of Soviet behavior." See W. I. Thomas, quoted in Urie Bronfenbrenner, "Allowing for Soviet Perceptions, in Roger Fisher, ed., *International Conflict and Behavioral Science* (New York: Basic Books, 1964), 166; Robert C. Tucker, *Politics as Leadership* (Columbia, Mo.: University of Missouri Press, 1981), 59; and T. H. Rigby and R. F. Miller, *Political and Administrative Aspects of the Scientific and Technical Revolution in the USSR* (Canberra: Australian National University Occasional Paper, No. 11, 1976), 1.

1. "The Scientific-Technological Revolution" and Political and Social Change

1. "*The neolithic revolution* (from the word *neolith*, the New Stone Age) was marked by the emergence of cropping, cattle breeding, the handicrafts, and trade. The specialized tool became the chief instrument of labor, serving as an extension of the human hand, the natural instrument of labor. The neolithic revolution laid the material groundwork for transition from the primitive communal formation to antagonistic class society.

The industrial revolution was marked by the start of machine production, which in the course of less than two centuries traveled the way from manufacture to assembly line. The industrial revolution laid the material groundwork for the genesis and development of the capitalist socioeconomic formation.

The scientific and technical revolution has entailed the formation of a comprehensive and dynamic system, 'science-technology-production,' whose major specific feature is that scientific development tends to outpace technical development, and the latter to outpace production, with science becoming a direct productive force. In the present epoch, industrial and scientific-and-technical production coexist and interact. The possibilities of machine production have yet to be exhausted, while the share of automated production is still fairly small." I. L. Andreev, *The Noncapitalist Way: Soviet Experience and the Liberated Countries* (Moscow: Progress Publishers, 1977), 39.

On the historical development of Marxist and Soviet theories of scientific, technological, and social change and on the differences between the NTR and the industrial revolution of the eighteenth and nineteenth centuries, see *Sovremennaia nauchno-tekhnicheskaia revoliutsiia—istoricheskoe issledovanie* (Moscow: Nauka, 1970); Julian Cooper, "The Concept of the Scientific and Technical Revolution in Soviet Theory," University of Birmingham, CREES Discussion Paper, 9 (1973); and E. M. Mirsky, "Science Studies in the USSR (History, Problems, Prospects)," *Science Studies*, 2 (1972), 281–94.

2. Shukhardin and Gukov, eds., *op. cit.*, 4. In a series of bibliographical essays, the editors of *Voprosy filosofii* (hereinafter *Vf*) identify what they consider the most important Soviet studies of the NTR during the heyday of detente: "Sotsial'no-filosofskie problemy nauchno-tekhnicheskoi revoliutsii (issledovaniia 1971–1975 gg.)," 2 (1976), 37–53. We have examined all of the books cited by the editors of *Vf* on two major topics: (1) the basic characteristics and significance of the contemporary NTR and (2) Soviet leadership and administration—including Politburo-level societal guidance, the "rationalization" of decision making and implementation, and the "scientific management" of research, development, and production.
3. See Frederic J. Fleron, Jr., ed., *Technology and Communist Culture: The Socio-Cultural Impact of Technology under Socialism* (New York: Praeger, 1977), especially Julian Cooper's essay, "The Scientific and Technical Revolution in Soviet Theory," and the editor's introductory and concluding chapters; and John Gurley, *Challenges to Capitalism: Marx, Lenin, and Mao*, 2d ed. (San Francisco, Calif.: San Francisco Book Co., 1980), especially Chapter 2.

The diagram immediately below is a Western summary of Marx's materialist conception of history, from John Gurley, *Challenges to Communism* (San Francisco: Freeman, 1982), 23.

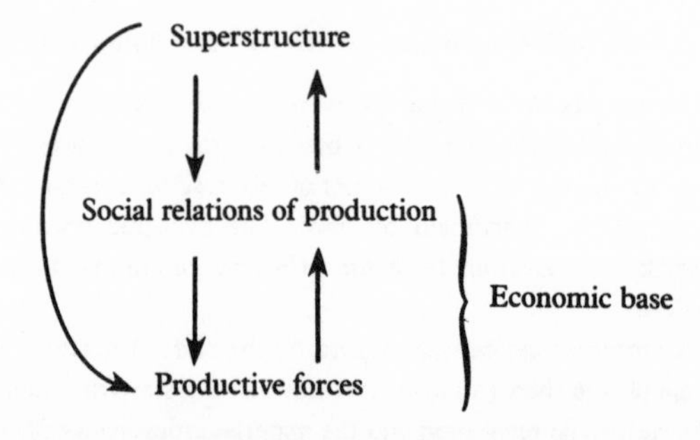

A Soviet comparison of the major elements of the productive forces and production relations is shown below. V. Tiagunenko, "Division of Labor and the Development of Social Production," *SS*, 4 (1975), 78.

Productive Forces	Boundary line for operation of laws	Relations of Production
(Sphere in which technological laws operate)		(Sphere in which economic laws operate)
Labor power as the capacity for work		Labor power as the capacity to create a surplus product
Machines		Use of machines
Technological division of labor		Social divisions of labor
Technical and technological division of instruments and means of production		Distribution of products in accordance with the forms of property in the means of production
Exchange of products		Exchange of commodities

4. *Partiia i sovremennaia nauchno-tekhnicheskaia revoliutsiia v SSSR*, 8–9 ff.
5. Cooper, "The Scientific and Technical Revolution in Soviet Theory," in Fleron, ed., *op. cit.*, 167.
6. See, e.g., Shukhardin and Gukov, eds., *op. cit.*, 144–47.
7. P. N. Fedoseev, in Ralf Dahrendorf et al., *Scientific-Technological Revolution: Social Aspects* (London and Beverly Hills, Calif.: Sage Publications, 1977), 88.
8. Shukhardin and Gukov, eds., *op. cit.*, 159–68.
9. A. G. Egorov, in E. M. Tiazhel'nikov, ed., *Za vysokoe kachestvo i deistvennost' ideologicheskoi raboty* (Moscow: Politizdat, 1981), 288–92 ff.
10. P. N. Fedoseev, *Dialektika sovremennoi epokhi*, 2d ed. (Moscow: Nauka, 1975). Fedoseev considerably expands his discussion of the NTR in the third edition of this book (Moscow: Nauka, 1978), especially 457–77.
11. *Partiia i sovremennaia nauchno-tekhnicheskaia revoliutsiia v SSSR*, 29–34.
12. *Ibid.*, 32–34; and N. V. Markov, *Nauchno-tekhnicheskaia revoliutsiia: analiz, perspektivy, posledstviia*, 2d ed. (Moscow: Politizdat, 1973), 198–229. Cf. C. Freeman's lucid conceptual distinctions in Ina Speigel-Rosing and Derek de Solla Price, eds., *Science, Technology, and Society: A Cross-Disciplinary Perspective* (Beverly Hills, Calif.: Sage Publications, 1977), 225–36. Freeman states, "It is important to distinguish between a technological change and a change in technique. . . . Whereas a technological change is an advance in knowledge, a change in technique is an alteration in the character of the equipment, products, and organization which are actually being used" (225).
13. Markov. *op. cit.*, 124.
14. *Ibid.*, 38–40, 124.
15. Julian Cooper, "The Scientific and Technical Revolution in the USSR" (paper prepared for presentation at the NASEES Annual Conference, Cambridge, England, 1981), 5–6.
16. *Ibid.*, 14–15.
17. P. N. Fedoseev, "Marxism-Leninism is the Firm Foundation of Soviet Social Science," in *Historical Materialism: Theory, Methodology, Problems* (Moscow: Social Sciences Today, 1977), 14.
18. V. G. Afanas'ev, "Scientific Management of Society," in *ibid.*, 62.
19. See, e.g., V. D. Kamaev, *Sovremennaia nauchno-tekhnicheskaia revoliutsiia: ekonomicheskie formy i zakonomernosti* (Moscow: Mysl', 1972), 54–55; and V. G. Marakhov, *Nauchno-tekhnicheskaia revoliutsiia i ee sotsial'nye posledstviia* (Moscow: Vysshaia shkola, 1975).

20. *Partiia i sovremennaia nauchno-tekhnicheskaia revoliutsiia v SSSR*, 44.
21. *Nauchno-tekhnicheskaia revoliutsiia i sotsializm*, 61 (emphasis in original).
22. P. N. Fedoseev, "Social Science and Progress," *SS*, 3 (1979), 21–22.
23. V. Cherkovets, "Production Relations under Developed Socialism," in *Economic Problems of Developed Socialism* (Moscow: Social Sciences Today, 1975), 80 (emphasis added).
24. V. V. Kosolapov, *Mankind and the Year 2000* (Moscow: Progress Publishers, 1976), 21–27 (emphasis added).
25. See Marakhov, *op. cit.*, 131–32 ff.
26. I. Gerasimov, "The Scientific and Technological Revolution and Geographical Science," in *Soviet Geographical Studies* (Moscow: Social Sciences Today, 1976), 5–6 (emphasis in original).
27. *Ibid.*, 6–10 (emphasis in original).
28. Iu. E. Volkov, "Vliianie nauchno-tekhnicheskoi revoliutsii na sistemu vlasti i demokraticheskie uchrezhdeniia," in *Sotsiologiia i sovremennost'*, Vol. 1 (Moscow: Nauka, 1977), 89.
29. G. N. Volkov, *Man and the Challenge of Technology* (Moscow: Novosti, 1972), 11.
30. See *USSR: Scientific and Technological Revolution* (Moscow: Novosti, 1973), 36–40 (chapter by G. N. Volkov).
31. *Ibid.*, 37. See also G. N. Volkov, *Istoki i gorizonty progressa: sotsiologicheskie problemy razvitiia nauki i tekhniki* (Moscow: Politizdat, 1976), 245 ff.
32. For elaboration of these and the following themes, see Fleron, ed., *op. cit.*
33. G. N. Volkov, *Man and the Challenge of Technology*, 37–38.
34. B. Liebson, *Partiinoe stroitel'stvo*, 1 (1946), cited in John Armstrong, *The Politics of Totalitarianism* (New York: Random House, 1961), 182.
35. V. Chikin, *Komsomolskaia pravda*, January 11, 1969, cited in Daniel Bell, *The Coming of Post-Industrial Society: A Venture in Social Forecasting* (New York: Basic Books, 1976), 354.
36. *Nauchno-tekhnicheskaia revoliutsiia i sotsial'nyi progress* (Moscow: Politizdat, 1972), 37.
37. L. I. Brezhnev, in *Materialy XXIV s'ezda KPSS* (Moscow: Politizdat, 1971), 57 (emphasis in original).
38. *Chelovek-nauka-tekhnika*, 322, 192; translation, 334–35, 204 (emphasis added).
39. See, e.g., the above quote from *Chelovek-nauka-tekhnika*; and A. S. Akhiezer, *Nauchno-tekhnicheskaia revoliutsiia i nekotorye sotsial'nye problemy proizvodstva i upravleniia* (Moscow: Nauka, 1974).
40. *Chelovek-nauka-tekhnika*, 324–25; translation, 336–37 (emphasis added).
41. Brezhnev's mainstream perspectives are emphasized in the present study. The archetypical reformer is Kosygin, and the archetypical conservative is Suslov. See A. N. Kosygin, *K veliki tseli: izbrannye rechi i stat'i*, 2 vols. (Moscow: Politizdat, 1979); and Kosygin, *K veliki tseli: o glavnom smysle zhizni—sluzhenii vysokim idealam partii* (Moscow: Molodaia gvardiia, 1981). Also, M. A. Suslov, *Marksizm-leninizm i sovremennaia epokha: izbrannye rechi i stat'i*, 3 vols. (Moscow: Politizdat, 1972, 1977, 1982).
42. See Barrington Moore, Jr., *Soviet Politics—The Dilemma of Power: The Role of Ideas in Social Change* (New York: Harper & Row, 1965).
43. Jan Triska, ed., *Soviet Communism: Rules and Programs* (San Francisco: Chandler, 1962), 41 ff.
44. K. U. Chernenko, "In the CPSU Central Committee," *Pravda*, April 26, 1984, 1;

translation, *Current Digest of the Soviet Press* (hereinafter *CDSP*), 17 (1984), 2.

45. V. G. Afanas'ev, *Nauchno-tekhnicheskaia revoliutsiia, upravlenie, obrazovanie* (Moscow: Politizdat, 1972), 237; translation, *The Scientific and Technological Revolution—Its Impact on Management and Education* (Moscow: Progress, 1975), 190.
46. On the role of ideology in politics—especially communist politics—see, e.g., Martin Seliger, *Ideology and Politics* (New York: Free Press, 1976); Franz Schurmann, *Ideology and Organization in Communist China*, 2d ed. (Berkeley: University of California Press, 1971); Alfred Meyer, "The Functions of Ideology in the Soviet Political System," *Soviet Studies*, 3 (January, 1966), 273–85; David Joravsky, "Soviet Ideology," *Soviet Studies*, 1 (July, 1966), 2–19.
47. John Michael Montias's apt phrase.
48. For example, Brezhnev noted at the 25th Party Congress that at the behest of top party and government organs "academic institutes, working together with ministries and departments," had prepared the draft of a "comprehensive program for scientific and technical progress and its social and economic effects for 1976–1990." He went on to say, "It is necessary to continue work on this program because it is an organic part of the current and long-term planning, providing the orientations (*orientiry*) without a knowledge of which the economy cannot be directed (*rukovodit'*) successfully." *Pravda*, February 25, 1976, 6. This program is contained in *Osnovnye napravleniia ekonomicheskogo i sotsial'nogo razvitiia SSSR na 1981–1985 gody i na period do 1990 goda* (Moscow: Politizdat, 1980).
49. Afanas'ev, *Nauchno-tekhnicheskaia revoliutsiia*, 245; translation, 198.
50. D. M. Gvishiani, *Organizatsiia i upravlenie*, 2d ed. (Moscow: Nauka, 1974), 127–28; translation, *Organization and Management: A Sociological Analysis of Western Theories* (Moscow: Progress, 1972), 124.
51. L. I. Brezhnev, *Pravda*, February 25, 1975, 6 (emphasis added). Also, see Linda Lubrano, *Soviet Sociology of Science* (Columbus, Ohio: American Association for the Advancement of Slavic Studies, 1976).
52. Vera Dunham brought this to our attention in an informative conversation on the treatment of the NTR in contemporary Soviet fiction.
53. *Vf*, 2 (1976), 38, 48, 51, 53; and 2 (1974), 9–10. The editorial board of *Vf* was more conservative after the mid-1970s than it had been in the recent past, and this shift strengthens our contention that the NTR literature is important. Echoing their reformist predecessors, the conservative editors repeatedly stressed the pragmatic value of social and philosophical studies of the NTR.
54. D. M. Gvishiani, "Nauchno-tekhnicheskaia revoliutsiia i sotsial'nyi progress," *Vf*, 4 (1974), 8.
55. Chernenko, in *CDSP*, 3 (emphasis added).
56. See, e.g., William Leiss, "The Social Consequences of Technological Progress: Critical Comments on Recent Theories," *Canadian Public Administration*, 3 (1970), 252, 260 ff.
57. This term is central to the analysis in Wendell Bell and James Mau, eds., *The Sociology of the Future: Theory, Cases, and Annotated Bibliography* (New York: Russell Sage Foundation, 1971). For a discussion of Soviet views of the future during the Khrushchev years, see Jerome Gilison, *The Soviet Image of Utopia* (Baltimore: Johns Hopkins University Press, 1975).
58. Fedoseev, "Social Science and Social Progress," 21, 29.

59. K. U. Chernenko, "Avangardnaia rol' partii kommunistov," *Kommunist*, 6 (1982), 30.
60. For an interpretation of "mediation," see Fleron's concluding chapter in Fleron, ed., *op. cit.* Cf. Radovan Richta in Dahrendorf et al., *op. cit.*, 25–72.

2. "Developed Socialism" and Progress

1. Under Brezhnev leading party officials and social theorists expounded at length on the idea of developed socialism, and they gave it a prominent place in the new USSR Constitution of 1977. See Robert Sharlet, ed., *The New Soviet Constitution of 1977: Analysis and Text* (Brunswick, Ohio: King's Court Communications, 1978), 75–76 ff. Among the major Soviet explications of the concept of developed socialism are: *Razvitoe sotsialisticheskoe obshchestvo: sushchnost', kriterii zrelosti, kritika revizionistskikh kontseptsii*, 2d ed., 3d ed. (Moscow: Mysl', 1975, 1979); *Sotsialisticheskoe obshchestvo: sotsial'no-filosofskie problemy sovremennogo sovetskogo obschchestva* (Moscow: Politizdat, 1975); V. I. Kas'ianenko, *Razvitoi sotsializm: istoriografiia i metodologiia problemy* (Moscow: Mysl', 1976).
2. P. N. Fedoseev, "Man and the Scientific-Technological Revolution," *SS*, 9 (1978), 32.
3. G. N. Volkov, "Izmenenie sotsial'noi orientatsii nauki," *Vf*, 1 (1969), 45.
4. A. I. Berg, "Science and Socialism," *World Marxist Review*, 11 (January, 1968), 1.
5. Kas'ianenko, *op. cit.*, 49.
6. A. G. Egorov, "Partiia nauchnoi kommunizma," *Kommunist*, 2 (1973), 43 (emphasis in original).
7. G. Kh. Shakhnazarov, *The Destiny of the World: The Socialist Shape of Things to Come* (Moscow: Progress, 1979), 136.
8. R. I. Kosolapov, *Sotsializm: k voprosam teorii*, 2d ed. (Moscow: Mysl', 1979), 403; translation, *Socialism: Questions of Theory* (Moscow: Progress, 1979), 355–56.
9. E. M. Babosov, "Scientific and Technological Revolution: The Growing Role of Scientific and Technological Intelligentsia," paper presented at the Eighth World Congress of Sociology, Toronto, August 17–24, 1974, 15–16.
10. N. V. Markov, "Trud, umstvennyi i fizicheskii," *Vf*, 11 (1968), 37.
11. *Razvitoe sotsialisticheskoe obschchestvo*, 2d ed., 93; and Iu. A. Levada, "Soznanie i upravlenie v obshchestvennykh protsessakh," *Vf*, 5 (1966), 62–73.
12. N. G. Nikolaev, "Rol' nauki v sotsialisticheskom obshchestve," *Vf*, 3 (1966), 12–13.
13. G. N. Volkov, "Izmenenie sotsial'noi orientatsii nauki," 43.
14. *Sotsialisticheskoe obshchestvo*, 310–21.
15. V. V. Semenov, "Novye iavleniia v sotsial'noi strukture sovetskogo obshchestva," in *XXIV s'ezd KPSS i problemy nauchnogo kommunizma* (Moscow: Politizdat, 1973), 55.
16. V. Kelle, "The Perfecting of Social Relations under Socialism," *Pravda*, April 5, 1967, 3–4; translation, *CDSP*, 19, (1967), 11.
17. Kas'ianenko, *op. cit.*, 124–29.
18. A. Bykov, "Socialism and the Scientific-Technological Revolution," *International Affairs* (hereinafter *IA*), 1 (1978), 39
19. *Razvitoe sotsialisticheskoe obshchestvo*, 2d ed., 163.
20. See, e.g., Volkov, *Istoki i gorizonty progressa*, 249–50.
21. Kas'ianenko, *op. cit.*
22. *Sotsialisticheskoe obshchestvo*, 6 (emphasis added); see also 128–66.
23. A very brief version of this section appeared in Erik P. Hoffmann, "Soviet Politics in

the 1980s," in Erik P. Hoffmann, ed., *The Soviet Union in the* 1980s (New York: The Academy of Political Science, 1984), 231–32.

24. Shakhnazarov, *The Destiny of the World*, 136. According to the USSR Constitution of 1977, "The basic direction of the development of the political system of Soviet society is the further unfolding of socialist democracy: the ever wider participation of citizens in the administration of the affairs of the state and of society, the improvement of the state apparatus, an increase in the activeness of public organizations, the intensification of people's control, the strengthening of the legal foundations of state and public life, greater publicity, and constant consideration for public opinion." Sharlet, ed., *op. cit.*, 79.
25. Sharlet, ed., *op. cit.*, 78. A. K. Belykh offers a detailed enumeration of the responsibilities and activities of the CPSU.

    "Thus, the functions of the party consist of the following:

    (1) the party is the leader of the entire political-management system of society;

    (2) the party works out the general line of development of society as a whole and the strategy and tactics of communist construction;

    (3) the party exercises political leadership in society, safeguarding the interests and ideals of the working class and its leading role, and strengthening and developing the union of the workers, peasants, and intelligentsia;

    (4) the party exercises general political leadership, safeguarding above all the public interests by combining them with group and individual interests, and also by directing the state and all organizations of the workers;

    (5) the party selects, assigns, and educates cadres in all organs of the political-management system;

    (6) the party is the chief political organizer of the population in fulfilling the tasks of communist construction;

    (7) the party is the chief political educator of the people;

    (8) the party harmoniously combines the national and international interests of the society's workers, following the principle of the indivisibility of the international and national tasks of a given socialist movement and of the world communist movement as a whole;

    (9) the party provides for the scientific-political leadership of society, spurring the creative and timely development of Marxist-Leninist theory." *Razvitoi sotsializm: sushchnost' i zakonomernosti* (Leningrad: Lenizdat, 1982), 170–71.
26. Kas'ianenko, *op. cit.*, 173.
27. G. Kh. Shakhnazarov, *Sotsialisticheskaia demokratiia: nekotorye voprosy teorii*, 2d ed. (Moscow: Politizdat, 1974), 68; translation, *Socialist Democracy: Aspects of Theory* (Moscow: Progress, 1974), 43–44.
28. *Ibid.*, 125; translation, 78.
29. E. M. Chekharin, *The Soviet Political System under Developed Socialism* (Moscow: Progress, 1977), 60.
30. E.g., Iu. A. Tikhomirov, ed., *Sovetskoe gosudarstvo v usloviiakh razvitogo sotsialisticheskogo obschchestva* (Moscow: Nauka, 1978), 207–29.
31. Kas'ianenko, *op. cit.*, 163.
32. B. I. Topornin, "NTR i sotsialisticheskoe gosudarstvo," in *Mezhdunarodnye otnosheniia, politika i lichnost'* (Moscow: Nauka, 1976), 50–51.
33. *Razvitoe sotsialisticheskoe obshchestvo*, 2d ed., 169.

34. Shakhnazarov, *The Destiny of the World*, 157–60.
35. Chekharin, *op. cit.*, 219.
36. *Sotsialisticheskoe obshchestvo*, 144 (emphasis added); and E. Kapustin, "The Scientific-Technological Revolution and the Improvement of Socialist Production Relations," *SS*, 1 (1975), 75–89.
37. See Chapters 3 and 4 for elaboration of this theme.
38. E.g., V. G. Marakhov and Iu. S. Meleshchenko, "Osobennosti i sotsial'nye posledstviia NTR," *Filosofskie nauki* (hereinafter *Fn*), 5 (1968), 93–100.
39. E.g., Ia. M. Zhukovskii, *Nauka kak proizvoditel'naia sila obshchestva* (Moscow: Mysl', 1973).
40. V. G. Marakhov, *Nauchno-tekhnicheskaia revoliutsiia i kommunizm* (Moscow: Znanie, 1971), 8.
41. S. P. Mal'skii, *Nauka i vsestoronnee razvitie lichnosti* (Kishinev: Shtinitsa, 1974), Chapter one.
42. Iu. S. Meleshchenko, *Tekhnika i zakonomernosti ee razvitiia* (Leningrad: Lenizdat, 1970), 122–34.
43. S. A. Tiushkevich, in N. A. Lomov, ed., *Nauchno-tekhnicheskii progress i revoliutsiia v voennom dele* (Moscow: Voenizdat, 1973), translated by the United States Air Force, U.S. Government Printing Office, 1973, 12.
44. *Nauchno-tekhnicheskaia revoliutsiia i sotsializm*, 21–29.
45. *Chelovek-nauka-tekhnika*, 40, 83–89; translation, 43, 84–89.
46. G. N. Volkov, *Istoki i gorizonty progressa*, 167.
47. *Ibid.*, 168.
48. *Ibid.*, 189.
49. *Ibid.*, 168.
50. *Ibid.*, 171–72.
51. E.g., "NTR i ee sotsial'naia problematika," *Vf*, 12 (1971), 3–16.
52. "Chelovek-nauka-tekhnika," *Vf*, 8 (1972), 33 (emphasis added).
53. J. K. Galbraith's apt phrase.
54. Karl Marx's apt phrase.
55. *The Political Economy of Capitalism* (Moscow: Progress, 1974), 222 (emphasis added).
56. For elaboration of Soviet thinking about advanced capitalism, East-West relations, and the NTR, see Hoffmann and Laird, *The Politics of Economic Modernization in the Soviet Union*, especially Chapters 7–10, and Hoffmann and Laird, *"The Scientific-Technological Revolution" and Soviet Foreign Policy*, especially Chapters 2–5. See also Robbin F. Laird, "Post-Industrial Society: East and West," *Survey*, 4 (Autumn, 1975), 1–17; and *idem*, "'Developed' Socialist Society and the Dialectics of Development and Legitimation in the Soviet Union," *Soviet Union*, 1 (1977), 130–49.
57. V. I. Gromeka, *Nauchno-tekhnicheskaia revoliutsiia i sovremennyi kapitalizm* (Moscow: Politizdat, 1976), 171.
58. E.g., I. A. Kozikov, *Problemy sootnosheniia nauchno-tekhnicheskoi i sotsial'noi revoliutsii* (Moscow: Izdatel'stvo MGU, 1972).
59. E.g., *Nauchno-tekhnicheskaia revoliutsiia i nekotorye problemy proizvodstva i upravleniia* (Moscow: Nauka, 1974), 103, 119 ff.
60. V. G. Marakhov, ed., *Soedinenie dostizhenii NTR s preimushchestvami sotsializme* (Moscow: Mysl', 1977), 68–74.
61. E. D. Modrzhinskii and Ts. A. Stepanian, eds., *The Future of Society* (Mos-

cow: Progress, 1973), Part four.

62. V. Turchenko, *The Scientific-Technological Revolution and the Revolution in Education* (Moscow: Progress, 1976), 68.
63. Gvishiani, *Organizatsiia i upravlenie*, 31–32; translation, 31.
64. *Ibid.*, 31; translation, 30.
65. *Ibid.*, 21; translation, 21.
66. See especially Fleron's introduction and conclusion in Fleron, ed., *op. cit.*
67. See Chapter 1.
68. See, e.g., Soviet critiques of Daniel Bell and J. K. Galbraith.
69. E.g., G. A. Arbatov, *Ideologicheskaia bor'ba v sovremennykh mezhdunarodnykh otnosheniiakh* (Moscow: Politizdat, 1970), especially the introduction; *Bor'ba idei* NTR (Leningrad: Lenizdat, 1973); V. Sol'shakov in *Komsomol'skaia pravda*, August 25 and 26, 1970; N. V. Pilipenko, "The Scientific-Technological Revolution and the Conception of a 'New Type of Society,'" in *Sociology and the Present Age* (Moscow: Soviet Sociological Association, 1974), 98.
70. I. Gerasimov, "Nazad k tekhnofobii," *Mirovaia ekonomika i mezhdunarodnye otnosheniia* (hereinafter *Memo*), 12 (1971), 132–34; and V. F. Kormer and Iu. P. Senokosov, "O tekhnologicheskogo determinizma k posttekhnokraticheskogo videniie," *Vf*, 7 (1973), 34–45; and K. I. Shilin, "Nravstvennyi krizis v amerikanskoi nauke," *Vf*, 11 (1971), 164–68; and G. S. Khozin, "Nauka i tekhnika, ideologiia i politika," *Vf*, 1 (1973), 71–82.
71. Pilipenko, *op. cit.*, 99–100 (emphasis added).
72. *The Fundamentals of Marxist-Leninist Philosophy*, 502.
73. See Chapter 1.
74. E.g., *Osnovnye napravleniia ekonomicheskogo i sotsial'nogo razvitiia SSSR na 1981–1985 gody i na period do 1990 goda.*

3. "The Scientific Management of Society": Decision-Making and Implementation Challenges

1. On the historical origins of issues raised in this chapter and for a discussion of the CPSU's role in a developed socialist society, see Erik P. Hoffmann, "Soviet Perspectives on Leadership and Administration," in Hoffmann and Laird, eds., *The Soviet Polity in the Modern Era*, 109–30.
2. Paul Cocks, "The Role of the Party in Soviet Science and Technology Policy," final report to the National Council for Soviet and East European Research, 1982, 10 ff.; and Paul Cocks, "Administrative Reform and Soviet Politics," *Soviet Economy in the 1980s: Problems and Prospects*, Vol. 1 (Washington, D.C.: U.S. Government Printing Office, 1983), passim.
3. See V. G. Afanas'ev, *Nauchnoe upravlenie obschchestvom*, 2d ed. (Moscow: Politizdat, 1973), 111–12, 266; Afanas'ev, *Nauchno-tekhnicheskaia revoliutsiia*; Afanas'ev, *Sotsial'naia informatsiia i upravlenie obschchestvom* (Moscow: Politizdat, 1975); translation, *Social Information and the Regulation of Social Development* (Moscow: Progress, 1978); Afanas'ev, *Chelovek v upravlenii obschchestvom* (Moscow: Politizdat, 1977).
4. *Ibid.*, 122.
5. *Ibid.*, 177–78 (emphasis added); also 120, 211.
6. *Ibid.*, 235.

7. Paul Cocks, "Rethinking the Organizational Weapon: The Soviet System in a Systems Age," *World Politics*, 2 (January, 1980), 253 (emphasis added).
8. V. Zaslavsky, "A New Phase in Soviet Sociology," *Radio Liberty Research Supplement*, April 9, 1976, 7.
9. B. Z. Mil'ner, in *Pravda*, June 1, 1979, 2; translation, *CDSP*, 22 (1979), 5 (emphasis added).
10. V. A. Trapeznikov, in *Pravda*, May 7, 1982, 2–3; translation, *CDSP*, 18 (1982), 3–4 (emphasis added).
11. A brief version of this section on "democratic centralism" appeared in Hoffmann, in Hoffmann, ed., *The Soviet Union in the* 1980s, 233–36.

    Article 3 of the 1977 USSR Constitution affirms: "The organization and activity of the Soviet state are constructed in accordance with the principle of democratic centralism: the elective nature of all bodies of state power, from top to bottom, their accountability to the people, and the binding nature of the decisions of higher bodies on lower. Democratic centralism combines united leadership with local initiative and creative activeness, with the responsibility of every state agency and official for the assigned task." In Sharlet, ed., *op. cit.*, 77.

    According to the CPSU rules, "The guiding principle of the organizational structure of the party is democratic centralism, which signifies:

    a. Election of all leading party bodies, from the lowest to the highest;
    b. Periodic reports of party bodies to their party organizations and to higher bodies;
    c. Strict party discipline and subordination of the minority to the majority;
    d. The decisions of higher bodies are obligatory for lower bodies." *Ustav Kommunisticheskoi Partii Sovetskogo Soiuza* (Moscow: Politizdat, 1973), III/19, 21–22.

    P. A. Rodionov provides historical perspective on the long-standing CPSU rule presented above. See *Printsip demokraticheskogo tsentralizma v stroitel'stve i deiatel'nosti Kommunisticheskoi partii* (Moscow: Politizdat, 1973), 44 ff.
12. *Ustav*, I/3/b, 11.
13. V. G. Afanas'ev, "Nauchnoe upravlenie obshchestvom i demokraticheskii tsentralizm," in *Sovetskaia demokratiia v period razvitogo sotsializma* (Moscow: Mysl', 1976), 152 (emphasis added); translation, *Soviet Democracy in the Period of Developed Socialism* (Moscow: Progress, 1979), 145.
14. G. Kh. Shakhnazarov, *Fiasko futurologii* (Moscow: Politizdat, 1979), 258; translation, *Futurology Fiasco* (Moscow: Progress, 1982), 166
15. R. I. Kosolapov, *Sotsializm*, 462 (emphasis added); translation, 409.
16. Shakhnazarov, *Fiasko futurologii*, 265 (emphasis added); translation, 171.
17. R. I. Kosolapov, *Sotsializm*, 458 (emphasis in original); translation, 406.
18. M. I. Piskotin, ed., *Problemy obshchei teorii sotsialisticheskogo gosudarstvennogo upravleniia* (Moscow: Nauka, 1981), 212–13.
19. *Ibid.*, 213 (emphasis added).
20. M. I. Piskotin, "Demokraticheskii tsentralizm; problemy sochetaniia tsentralizatsii i detsentralizatsii," *Sgp*, 5 (May, 1981), 40.
21. Shakhnazarov, *Fiasko futurologii*, 263–65 (emphasis added); translation, 168–71. The indented statement is by N. Iribadiakov, a Bulgarian philosopher whom Shakhnazarov cites with praise.
22. *Ibid.*, 171–75.
23. *Ibid.* (emphasis added).

24. R. I. Kosolapov, *Sotsializm*, 457–58 ff. (emphasis added); translation, 405–6 ff.
25. Sharlet, ed., *op. cit.*, 92–93.
26. Afanas'ev, in *Sovetskaia demokratiia*, 145, 148–49; translation, 140, 143–44.
27. E.g., N. A. Tsagolov, ed., *Nauchno-tekhnicheskaia revoliutsiia i sistema ekonomicheskikh otnosheniia razvitogo sotsializma* (Moscow: Izdatel'stvo MGU, 1979), 56–57, 83–84; B. Z. Mil'ner, "Organization of the Management of Production," *SS*, 3 (1976), 51–55; M. I. Piskotin, "Centralism and Democratic Principles," *SS*, 4 (1982), 55–67.
28. Afanas'ev, in *Sovetskaia demokratiia*, 149; translation, 144. Shakhnazarov, in *ibid.*, 166; translation, 158. Also *Fiasko futurologii*, 271; translation, 175. And see note 23.
29. Piskotin, *Problemy obshchei teorii*, 213. Afanas'ev in *Sovetskaia demokratiia*, 147; translation, 142.
30. Mil'ner, "Organization of the Management of Production," 53; Piskotin, "Centralism and Democratic Principles," 63.
31. E.g., A. Vodolazskii, "Vysshii printsip partiinogo rukovodstva," *Kommunist*, 12 (1979), 38 ff.
32. E.g., A. P. Butenko, *Politicheskaia organizatsiia obshchestva pri sotsializme* (Moscow: Mysl', 1981), 95–96, 169, 193, 203 ff.
33. E.g., see the chapters by Burlatskii, Butenko, and Fedoseev in *Lenin kak politicheskii myslitel'* (Moscow: Politizdat, 1981).
34. A. P. Butenko, "Eshche raz o protivorechiiakh sotsializma," *Vf*, 2 (1984), 124–29; translation, *CDSP*, 24 (1984), 6–7 (first two emphases added, third in original). See also Butenko's essay in *Vf*, 10 (1982), 16–29.
35. E.g., Butenko, *Politicheskaia organizatsiia obshchestva*, 169, 193 ff. (emphasis added).
36. *Ibid.*, 189–90 (emphasis added). Also, see Butenko's overview essay on the concept of developed socialism in the collective Soviet–East European volume he edited. *Razvitoi sotsializm: obshchee i spetsificheskoe v ego stroitel'stve* (Moscow: Nauka, 1980), 35–63. To ensure wider distribution of Butenko's essay, a shortened version was published as the lead article in *Obn*, 1 (1981), 5–18.
37. Mil'ner, "Organization of the Management of Production," 52.
38. Butenko, *Politicheskaia organizatsiia obshchestva*, 183 (emphasis added).
39. L. I. Brezhnev, in *Current Soviet Policies VIII*, 29.
40. Ustav, I/3/b, 11; III/26, 26–27; IV/35, 32; V/42/f, 38.
41. Alfred Meyer, *Leninism* (Cambridge, Mass.: Harvard University Press, 1957), 95.
42. Brezhnev, in *Current Soviet Policies VIII*, 28.
43. *Ibid.* (emphasis added).
44. Afanas'ev, *Nauchnoe upravlenie obshchestvom*, 1st ed. (1968), 109: V. V. Semenov, "Issledovanie nauchnogo upravlenie obchchestvom," *Kommunist*, 8 (1968), 128; V. G. Afanas'ev, ed., *Nauchnoe upravlenie obshchestvom*, Vol. 3 (Moscow: Mysl', 1969), 24–25 ff. (emphasis in original). We are indebted to David Holloway for bringing this exchange to our attention.
45. Afanas'ev, in *Sovetskaia demokratiia*, 136 (emphasis added); translation, 133.
46. F. M. Burlatskii, *The Modern State and Politics* (Moscow: Progress, 1978), 88; F. M. Burlatskii, *Lenin, gosudarstvo, politika* (Moscow: Nauka, 1970), 84 ff.
47. N. V. Chernogolovkin, in *Sovetskaia demokratiia v period razvitogo sotsializma*, 43; translation, 54.
48. Afanas'ev, in *ibid.*, 136; translation, 133. See also R. V. Martanus, *Nauchnoe rukovodstvo i upravlenie sotsialisticheskim obshchestvom* (Moscow: Izdatel'stvo MGU,

1978), 12–13 ff.

49. Afanas'ev, *Sotsial'naia informatsiia*, 128; translation, 132. See also *Upravlenie sotsialisticheskim proizvodstvom: voprosy teorii i praktiki*, 3d ed. (Moscow: Ekonomika, 1978); *Gosurdarstvennoe upravlenie v SSSR v usloviiakh nauchno-tekhnicheskoi revoliutsii* (Moscow: Nauka, 1978); and N. A. Tsagolov, ed., *Nauchno-tekhnicheskaia revoliutsiia i sistema ekonomicheskikh otnoshenii razvitogo sotsializma* (Moscow: Izdatel'stvo MGU, 1979).
50. Iu. A. Krasin, "Some Questions of the Methodology of Political Thinking," in *Time, Space, and Politics*, 45 (emphasis added).
51. K. Varmalov, *Socialist Management: The Leninist Concept* (Moscow: Progress, 1977), 297–98.
52. E.g., A. G. Vendelin, *Podgotovka i priniatie upravlenicheskogo resheniia* (Moscow: Ekonomika, 1977); Vendelin, *Protsess priniatiia reshenii* (Tallin: Valgus, 1973).
53. Afanas'ev, *Sotsial'naia informatsiia*, 155; translation, 159–60. See also Afanas'ev, *Chelovek v upravlenii obshchestvom*, especially Chapter 3. For further analysis and specifics concerning the informational aspects of decision making and implementation in the USSR, see Erik. P. Hoffmann, "Information Processing the Party: Recent Theory and Experience," in Karl Ryavec, ed., *Soviet Society and the Communist Party* (Amherst, Mass.: University of Massachusetts Press, 1978), 63–87.
54. Afanas'ev, *Chelovek v upravlenii obshchestvom*, 96 ff.
55. Stephen Robbins, "Reconciling Management Theory with Management Practice," *Business Horizons*, 1 (February, 1977), 39: quoted in V. G. Afanas'ev and B. Z. Mil'ner, "Amerikanskie burzhuaznye teorii upravleniia," *Obshchestvennye nauki* (hereinafter, *Obn*), 3 (1979), 141.
56. Burlatskii, *The Modern State and Politics*, 141; and Burlatskii, *Lenin, gosudarstvo, politika*, passim.
57. Burlatskii, *the Modern State and Politics*, 141–48.
58. *Ibid.*, 111 ff.
59. K. U. Chernenko, "Leninskaia strategiia rukovodstva," *Kommunist*, 13 (1981), 13.
60. I. V. Kapitonov, in Tiazhel'nikov, ed., *Za vysokye kachestvo i deistvennost' ideologicheskoi raboty*, 223–24.
61. L. I. Brezhnev, *Voprosy razvitiia politicheskoi sistemy sovetskogo obshchestva* (Moscow: Politizdat, 1977), 161 (emphasis added).
62. E. P. Kireev, ed., *Rukovodiashchiia i organizuiushchiia rol' KPSS v period razvitogo sotsializma* (Leningrad: Izdatel'stvo Leningradskogo universiteta, 1978), 44.
63. *Ibid.*, 44–46.
64. A statement by A. E. Bovin in 1968, quoted in *ibid.*, 45 (emphasis added).
65. *Ibid.*, 53.
66. *Ibid.*, 48, 54, 57 ff.
67. For elaboration, see Erik P. Hoffmann, "The 'Scientific' Management of Soviet Society," *Problems of Communism*, 36 (May–June, 1977), 59–67.
68. Quoted by Krasin, in *Time, Space, and Politics*, 51.
69. V. M. Glushkov, in *Nauchno-tekhnicheskoe obshchestvo v SSSR* (January, 1972), 1; translation, Foreign Broadcast Information Service, *Cybernetics in the USSR* (Washington, D.C.), 111, June 30, 1972, 1.
70. Akhiezer, *Nauchno-tekhnicheskaia revoliutsiia i nekotorye problemy proizvodstva i upravleniia*, 6–7.

71. Mil'ner, in *Pravda*, June 1, 1979, 2; translation, *CDSP*, 22 (1979), 5.
72. *Upravlenie sotsialisticheskim proizvodstvom: voprosy teorii i praktiki*, 2d ed. (Moscow: Ekonomika, 1975), 431.
73. *Ibid.*, 166, 429–33.
74. D. M. Gvishiani, "The Philosophical Basis of Systems Studies," *SS*, 3 (1982), 62–65 (emphasis added).
75. See, e.g., Afanas'ev, *Nauchnoe upravlenie obshchestvom*, 1st ed., passim.
76. I. V. Blauberg, "The History of Science and the Systems Approach," *SS*, 3 (1977), 90–100.
77. *The Fundamentals of Marxist-Leninist Philosophy*, 151.
78. Afanas'ev, *Sotsial'naia informatsiia*, 33; translation, 38.
79. Afanas'ev, *Nauchno-tekhnicheskaia revoliutsiia*, 217; translation, 173.
80. See Chapter 5 and Conclusion.
81. I. V. Blauberg, V. N. Sadovskii, and E. G. Iudin, *Systems Theory: Philosophical and Methodological Problems* (Moscow: Progress, 1977), 262.
82. *Ibid.*, 267.
83. *Ibid.*
84. *Ibid.*, 284–86.
85. Gvishiani, "The Philosophical Basis of Systems Studies," 67.
86. V. G. Afanas'ev, "Systems Approach in Social Cognition," *SS*, 1 (1979), 44.
87. Afanas'ev, *Sotsial'naia informatsiia*, 115; not in the translation.
88. *Ibid.*, 154; translation, 158–59.
89. Afanas'ev, "Systems Approach in Social Cognition," 41 (emphasis added).
90. V. G. Afanas'ev, "Programmno-tselovoe planirovanie i upravlenie," *Obn*, 4 (1982), 19 ff.; for elaboration, see Afanas'ev, *Obshchestvo: sistemnost', poznanie i upravlenie* (Moscow: Politizdat, 1980).
91. V. M. Glushkov, "Problemy sotsial'no-ekonomicheskogo upravleniia v epokhu NTR," *Obn*, 2 (1982), 70 ff.
92. *Ibid.*, 65–66 (emphasis in original).
93. See Chapter 4.
94. B. Z. Mil'ner, "Printsipy formirovaniia organizatsionnykh struktur," *Obn*, 5 (1981), 104–11; for elaboration, see Mil'ner, *Organizatsiia programmno-tselovogo upravleniia* (Moscow: Nauka, 1980).
95. Mil'ner, "Printsipy," 112–13.
96. Afanas'ev, "Programmno-tselovoe planirovanie i upravlenie," 31.
97. *Ibid.*
98. *Ibid.*, 18 (emphasis added).
99. Iu. Sheinin, *Science Policy* (Moscow: Progress, 1978), 35, 214–15.
100. *Ibid.*, 289–90.
101. L. K. Naumenko, "Integratsiia nauk i povyshenie potentsiala nauchnogo tvorchestva," in *Nauchno-tekhnicheskaia revoliutsiia i stroitel'stvo kommunizma* (Moscow: Mysl', 1976), 157.
102. Afanas'ev, *Nauchno-tekhnicheskaia revoliutsiia*, 277; translation, 228
103. *Ibid.*, 278; translation, 228.
104. *Ibid.*, 280; translation, 230.
105. *Ibid.*, 286–287; translation, 235.
106. Cf. V. F. Turchin, *The Phenomenon of Science: A Cybernetic Approach to Human Evolu-*

tion (New York: Columbia University Press, 1977); and Mark Popovsky, *Manipulated Science: The Crisis of Science and Scientists in the Soviet Union Today* (Garden City, N.Y.: Doubleday, 1979).

107. P. A. Rachkov, *Naukovedenie* (Moscow: Izdatel'stvo MGU, 1974), 19–20.
108. S. R. Mikulinskii and N. I. Rodnyi, "Nauka kak predmet spetsial'nogo issledovaniia," *Vf*, 5 (1966), 29.
109. Rachkov, *Naukovedenie*, 149 ff.
110. V. A. Frolov, in S. R. Mikulinskii and M. G. Iaroshevskii, eds., *Nauchnoe tvorchestvo* (Moscow: Nauka, 1969), 271–80.
111. On some of the above issues, see, e.g., V. Zh. Kelle and S. R. Mikulinskii, eds., *Sotsiologicheskie problemy nauki* (Moscow: Nauka, 1974).
112. D. M. Gvishiani, in *Osnovnye printsipy i obshchie problemy upravleniia naukoi* (Moscow: Nauka, 1973), 40.
113. S. R. Mikulinskii, in E. A. Beliaev et al., eds., *Organizatsiia nauchnoi deiatel'nosti* (Moscow: Nauka, 1968), 139–40.
114. See, e.g., G. M. Dobrov, *Nauka o nauke: vvedenie v obshchee naukoznanie* (Kiev: Naukova dumka, 1970).
115. G. M. Dobrov, *Pravda*, October 15, 1970, 2; translation, *CDSP* 41 (1970), 11.
116. L. Gliazer, "Ekonomika nauk i nauka ekonomiki," *Vf*, 6 (1973); translation, *Problems of Economics* (December, 1973), 36.
117. *Ibid.*, 51–52.
118. S. R. Mikulinskii, *Kommunist*, 5 (1973); translation, *CDSP*, 21 (1973), 5.
119. *Ibid.*
120. Gvishiani, *Osnovnye printsipy*, 42 ff.
121. V. Zh. Kelle and S. R. Mikulinskii, "Sociology of Science," *SS*, 3 (1977), 86.
122. Gvishiani, *Osnovnye printsipy*, 43.
123. D. M. Gvishiani, in Kelle and Mikulinskii, eds., *Sotsiologicheskie problemy nauki*, 180.
124. See, e.g., Iakov Rabkin, "Measuring Science: Uses and Expectations," *Survey*, 2 (Spring, 1976), 79.
125. Rachkov, *Naukovedenie*, 131–32 (emphasis added).
126. See, e.g., I. I. Leiman, *Nauka kak sotsial'nyi institut* (Leningrad: Nauka, 1971).
127. A. Titmonas, in Kelle and Mikulinskii, eds., *Sotsiologicheskie nauki*, 160–61.
128. Paul Cocks, *Science Policy in the Soviet Union*, Vol. 2 of *Science Policy: USA/USSR* (Washington, D.C.: U.S. Government Printing Office, 1980), 5.
129. *Ibid.*, 13.
130. See, e.g., Cocks, *Science Policy in the Soviet Union*; Amann and Cooper, eds., *Industrial Innovation in the Soviet Union*; John Thomas and Ursula Kruse-Vaucienne, eds., *Soviet Science and Technology: Domestic and Foreign Perspectives* (Washington, D.C.: George Washington University Press, 1977); Berliner, *The Innovation Decision in Soviet Industry*; Eugene Zaleski et al., *Science Policy in the USSR* (Paris: OECD, 1969).
131. Philip Hanson, "Organizational Changes for Soviet Research and Development?" *Radio Liberty Research*, RL 464/82, November 10, 1982, 3, citing *Pravda*, August 1, 1982 (emphasis added).
132. V. G. Kostiuk, in *Nauka, organizatsiia i upravlenie* (Novosibirsk: Nauka, 1979), 98 ff.
133. *Ibid.*, 110, 117.

134. *Ibid.*, 110–11.
135. V. A. Rassokhin, in *Pravo i upravlenie nauchnymi organizatsiiami* (Moscow: Nauka, 1980), 79, 82 (emphasis in original).
136. Kostiuk, *op. cit.*, 117.
137. *Ibid.*, 117–20.
138. V. P. Rassokhin, in *Ekonomika i organizatsiia promyshlennogo proivzodstva* (hereinafter *Eko*), 1 (1980); translation, *CDSP*, 12 (1980), 3.
139. *Ibid.* (emphasis added).
140. Rassokhin, in *Pravo i upravlenie nauchnymi organizatsiiami*, 84 ff.
141. Rassokhin, in *Eko*.
142. Hanson, "Organizational Changes for Soviet Research and Development?" 2.
143. Exceptions include *Partiia i sovremennaia nauchno-tekhnicheskaia revoliutsiia v SSSR*; and Cocks, "The Role of the Party."
144. Cocks, "The Role of the Party," iii.
145. R. O. Khalfina, ed., *Sotsialisticheskoe pravo i nauchno-tekhnicheskaia revoliutsiia* (Moscow: Nauka, 1979), 23.
146. Burlatskii, *The Modern State and Politics*, 125 (emphasis added).
147. S. Kurits, *Pravda*, March 1, 1983, 2; translation, *CDSP*, 9 (1983), 7.
148. For elaboration, see Erik P. Hoffmann, "Socialist Perspectives on the 'Scientific and Technical Revolution,' Management, and Law," in Gordon Smith, Peter Maggs, George Ginsburgs, eds., *Soviet and East European Law and the Scientific-Technical Revolution* (Elmsford, N.Y.: Pergamon Press, 1981), 19–46, and Erik P. Hoffmann and Robbin F. Laird, "Soviet Economic Development and the World Economy: Legal Aspects of East-West Economic Relations," in Peter Maggs, Gordon Smith, George Ginsburgs, eds., *Law and Economic Development in the Soviet Union* (Boulder, Colo.: Westview, 1982), 257–90.
149. For elaboration of the themes in this section, see Erik P. Hoffmann, "Technology, Values, and Political Power in the Soviet Union: Do Computers Matter?" in Fleron, ed., *op. cit.*, 397–436.
150. From Afanas'ev's introduction to V. M. Grigorov, *Eksperty v sisteme upravleniia obshchestvennym proizvodstvom* (Moscow Mysl', 1976), 4.
151. P. N. Fedoseev, in *XXV s'ezd KPSS i zadachi kafedr obshchestvennykh nauk* (Moscow: Politizdat, 1977), 125.
152. See, e.g., R. G. Ianovskii, in *Izvestiia Sibirskogo Otdeleniia Akademii Nauk SSSR, seriia obshchestvennykh nauk*, 11 (1976), 5 ff.
153. See, e.g., P. V. Alekseev and A. Ia. Il'in, *Printsip partiinosti i estestvoznanie* (Moscow: Izdatel'stvo MGU, 1972), 30–34.
154. Afanas'ev, *Nauchno-tekhnicheskaia revoliutsiia*, 234–45; translation, 196–98.
155. See, e.g., James Danziger et al., *Computers and Political Power* (New York: Columbia University Press, 1982); and Kenneth Laudon, *Computers and Bureaucratic Reform* (New York, Wiley, 1974).
156. Steven Marcus, *The New York Times*, August 19, 1983, D3.
157. V. M. Grigorov, in *Nauchno-tekhnicheskaia revoliutsiia i stroitel'stvo kommunizma* (Moscow: Mysl', 1976), 177; Grigorov, *Eksperty*, 15–16 (emphasis added).
158. *Ibid.*, 80.
159. *Ibid.*, 70; and V. M. Grigorov, in *Nauchnoe upravlenie obshchestvom*, 6 (1972), 332–56.
160. G. M. Dobrov and L. Smirnov, *Izvestiia*, December 19, 1970, 2.

161. See, e.g., V. Shliapentokh, *Kak segodniia izuchaut zavtra* (Moscow: Mysl', 1975).
162. Afanas'ev, *Nauchno-tekhnicheskaia revoliutsiia*, 314–15; translation, 253 (emphasis added).
163. Shakhnazarov, in *Sovetskaia demokratiia v period razvitogo sotsializma*, 156; translation, 149.
164. Seymour Goodman, "Soviet Computing and Technology Transfer: An Overview," *World Politics*, 4 (July 1979), 567.
165. Aron Katsenelinboigen, *Soviet Economic Thought and Political Power in the USSR* (Elmsford, N.Y.: Pergamon, 1980), 147–48.
166. N. C. David and S. E. Goodman, "The Soviet Bloc's Unified System of Computers," *Computing Surveys*, 2 (June 1978), 111.
167. Glushkov, "Problemy sotsial'no-ekonomicheskogo upravlenie v epokhu NTR," 74–75.
168. *Ibid.*, 64.
169. Allan Kroncher, "Death of Academician Glushkov," *Radio Liberty Research*, RL 57/82, February 3, 1982, 2.
170. *Ibid.*; and *Pravda*, March 13, 1978.
171. Goodman, *op. cit.*, 569–70.
172. The case for considerable change is made by Loren Graham, "Science and Computers in Soviet Society," in Hoffmann, ed., *The Soviet Union in the 1980s*, 124–34.
173. Afanas'ev, *Sotsial'naia informatsiia*, 307 (emphasis added); translation, 259.
174. For elaboration, see Hoffmann and Laird, *The Politics of Economic Modernization in the Soviet Union* and *"The Scientific-Technological Revolution" and Soviet Foreign Policy*.
175. For a Kremlinological interpretation, see Christian Duevel, *Radio Liberty Research*, RL 375/74, November 6, 1974; and RL 5/75, December 20, 1974.
176. V. V. Varchuk and V. I. Razin, in *Vf*, 4 (1967), 142 ff.

4. The Mass Media, Communication Systems, and Dissent

1. Ned Temko, "Soviet Insiders: How Power Flows in Moscow," in Hoffmann and Laird, eds., *op. cit.*, 179.
2. Afanas'ev, *Sotsial'naia informatsiia*, 383; translation, 334–35 (emphasis in original).
3. Joseph Stalin, quoted in Isaac Deutscher, *Stalin: A Political Biography*, 2d ed. (New York: Oxford University Press, 1966), 328.
4. Isaiah Berlin, *Four Essays on Liberty* (New York: Oxford University Press, 1969), 132–33.
5. See, e.g., Wilbur Schramm, in Fred Siebert et al., *Four Theories of the Press* (Urbana, Ill.: University of Illinois Press, 1956), 121.
6. "Sredstva massovoi informatsii i propagandy," in *Nauchnyi kommunizm: slovar'*, 2d ed. (Moscow: Politizdat, 1975), 360 (first emphasis added; second in original).
7. N. G. Bogdanov and B. A. Viazemskii, *Spravochnik zhurnalista* (Leningrad: Lenizdat, 1965), 37–49.
8. See, e.g., M. N. Marchenko, "Politicheskaia organizatsiia sovetskogo obshchestva," *Sgp*, 4 (1976), 141–45.
9. See, e.g., P. N. Fedoseev, "Man and the Scientific and Technological Revolution," *SS*, 1 (1978), 32 ff.
10. V. M. Gorokhov and V. D. Pel't, eds., *Masterstvo zhurnalista* (Moscow: Izdatel'stvo MGU, 1977), 159–61.

11. *Fundamentals of Marxist-Leninist Philosophy*, 502.
12. Ellen Mickiewicz, *Media and the Russian Public* (New York: Praeger, 1981), vii.
13. Timothy Colton, *The Dilemma of Reform in the Soviet Union* (New York: Council on Foreign Relations, 1984), 7. Cf. Lilita Dzirkals et al., *The Media and Intra-Elite Communication in the USSR* (Santa Monica, Calif.: Rand, 1982).
14. Mickiewicz, *op. cit.*, passim. Also, Ellen Mickiewicz, "Policy Issues in the Soviet Media System," in Hoffmann, ed., *The Soviet Union in the 1980s*, 115 ff. (emphasis added); *idem*, "Feedback, Surveys, and Communication Theory," *Journal of Communication*, 2 (Spring 1983), 97–110; and *idem*, "Evaluation Studies of Soviet Party Members," *Public Opinion Quarterly*, 4 (Winter 1976–77), 482–84 ff.
15. Ellen Mickiewicz, "Policy Applications of Public Opinion Research in the Soviet Union," *Public Opinion Quarterly*, 4 (Winter 1972–73), 578.
16. V. Davydchenkov and L. Karpinskii, "The Reader and the Newspaper," *Izvestiia*, July 11, 1968, 5; translation, *CDSP*, 28 (1968), 18 (emphasis added).
17. S. Tsukasov, "The Soviet Press—Ways to Improve Effectiveness and Quality," *Kommunist*, 7 (1977), 15 ff.; translation, *CDSP*, 21 (1977), 13 ff.
18. M. V. Shkondin, in *Zhurnalistika v politicheskoi strukture obshchestva* (Moscow: Izdatel'stvo MGU, 1975), 52.
19. Zhores Medvedev, *The Medvedev Papers* (London: Macmillan, 1978), 259.
20. L. G. Churchward, *The Soviet Intelligentsia* (London: Routledge & Kegan Paul, 1973), 58.
21. *Annual Report* (Washington, D.C.: 1983), The Board for International Broadcasting, 19–20.
22. Quoted in Zev Katz, *The Communications System in the USSR* (Cambridge, Mass.: MIT Center for International Studies, 1977), 19.
23. Mickiewicz, *Media and the Russian Public*, 18; and *idem*, "Policy Issues in the Soviet Media," 113 ff.
24. Julian Cooper, "Is There a Technological Gap between East and West?" Paper prepared for Canadian Institute of International Affairs conference, June 1984, 3, 19.
25. R. A. Safarov, "Problemy issledovaniia obshchestvennogo mneniia," *Vf*, 1 (1977), 43 (emphasis added). See also *idem*, *Obshchestvennoe mnenie i gosudarstvennoe upravlenie* (Moscow: Iuridicheskaia literatura, 1975).
26. K. U. Chernenko, address to CPSU Central Committee plenum, *Pravda*, June 1, 1983, 1; translation, *CDSP*, 24 (1983), 10.
27. B. M. Firsov and K. Muzdybaev, "Constructing a System of Indicators of the Mass Media Audience," *Sotsiologicheskie issledovaniia*, 1 (1975), 113–20; translation, *Soviet Law and Government*, 4 (Spring 1976), 57.
28. See, e.g., M. T. Iovchuk and L. N. Kogan, eds., *The Cultural Life of the Soviet Worker* (Moscow: Progress, 1975), 173–85.
29. Anonymous, "The Two Systems in Action," *Comparative Studies in History*, 2 (April 1978), 248.
30. See Parrott, *op. cit.* Also, Hoffmann and Laird, *The Politics of Economic Modernization in the Soviet Union*, and *idem.*, *"The Scientific-Technological Revolution" and Soviet Foreign Policy.*
31. K. Beliak, in *Kommunist*, 8 (1983), responding to G. Popov, in *Kommunist*, 18 (1982); translation, *CDSP*, 23 (1983), 4–5.
32. Cynthia Enloe, *The Politics of Pollution in a Comparative Perspective* (New York:

McKay, 1975), 209.

33. See, e.g., the numerous responses in *Pravda* to G. Kol'bin, "Vremia i stil' raboty," *Pravda*, February 6, 1967, 2–3; translation, *CDSP*, 6 (1967), 23–24.
34. Lev Lifshitz-Losev, "What It Means to be Censored," *New York Review of Books*, 25 (June 29, 1978), 45.
35. Zhores Medvedev, *Soviet Science* (New York: Norton, 1978), 190–91 (emphasis in original).
36. See, e.g., Karen Winkler, "Paucity of Data about Soviet Publishing Concerns Western Librarians," *The Chronicle of Higher Education*, July 5, 1984, 7.
37. *Ibid.*, 217, 126 (emphasis in original).
38. Quoted in A. Romanov, "Journalism—A Most Important Province of Party and Public Activity," *Partiinaia zhizn'*, 9 (1961), 17–24; translation, *CDSP*, 17 (1961), 10.
39. M. Buzhkevich, "Obrashchaiutsia liudi v obkom," *Pravda*, August 9, 1967, 2.
40. Butenko, *Politicheskaia organizatsiia obshchestva pri sotsializme*, 182–93 (emphasis added).
41. Shakhnazarov, *Sotsialisticheskaia demokratiia*, 125–30; translation, 78–81.
42. See, e.g., G. T. Zhuravlev, *Sotsial'naia informatsiia i upravlenie ideologicheskim protsessom* (Moscow: Mysl', 1973), 30, 81 ff.
43. R. A. Safarov, "Public Opinion in the Conditions of Developed Socialism," *Kommunist*, 12 (1977), 29–40; translation, *CDSP*, 36 (1977), 5.
44. Safarov, "Informing the Population about the Activity of State Agencies," 31.
45. Stephen White, "Political Communications in the USSR: Letters to Party, State, and Press," *Political Studies*, 1 (March 1983), 46–57.
46. "8000 Directors," *Literaturnaia gazeta*, October 12, 1977, 12; translation, *CDSP*, 41 (1977), 22.
47. See, e.g., E. Klopov, Man after Work (Moscow: Progress, 1975), 120.
48. M. Odinets, "The Viewer Asks Questions," *Pravda*, December 26, 1974, 3; translation, *CDSP*, 52 (1974), 27 (emphasis added).
49. *Ibid.*, 26–27.
50. L. I. Lopatnikov, in V. D. Pel't and M. V. Shkondin, eds., *Problematika gazetnykh vystuplenii* (Moscow: Izdatel'stvo MGU, 1975), 13–39.
51. V. Seliunin, "Do We Write Properly about Laggard Enterprises?" *Zhurnalist*, 1 (1977), 51–53; translation, *CDSP*, 8 (1977), 7.
52. A. Rubinov, "Interview without Answers," *Literaturnaia gazeta*, March 22, 1978, 11; translation, *CDSP*, 11 (1978), 11.
53. Tsukasov, 14; translation, 13.
54. Theodore Pirard, "Soviet Television during the Cosmic Era," *Space World* (August–September 1978), 17; and *Jane's Spaceflight Directory* (London, 1984).
55. Mickiewicz, *Media and the Russian Public*, 19.
56. Editorial, "Television Facilities," *Pravda*, October 23, 1976, 1; translation, *CDSP*, 43 (1976), 34.
57. P. Barashev, "Lines Fly through Space," *Pravda*, January 14, 1977, 6; translation, *CDSP*, 2 (1977), 27.
58. Mickiewicz, *Media and the Russian Public*, 19 (emphasis in original).
59. I. G. Petrov and A. S. Seregin, in *Problemy nauchnogo kommunizma*, Vol. 5 (Moscow: Mysl', 1971), 316–53; translation, *Soviet Law and Government*, 4 (Spring 1973), 325.

60. R. A. Safarov, "Informing the Population about the Activity of State Agencies," *Sgp*, 2 (1974), 22–30; translation, *Soviet Law and Government*, 3 (Winter 1974–75), 40.
61. "O merakh po uluchsheniiu podgotovki i perepodgotovki zhurnalistskikh kadrov," Central Committee decree, January 20, 1975, in *Ob ideologicheskoi rabote KPSS: sbornik dokumentov* (Moscow: Politizdat, 1977), 629.
62. Tsukasov, 21; translation, 14.
63. A. Iurovskii, "The Journalist Speaks," *Pravda*, December 19, 1969, 3; translation, *CDSP*, 51 (1969), 36.
64. Tsukasov, 17–18; translation, 13 (emphasis added).
65. White, *op. cit.*, 54 (emphasis added).
66. Tsukasov, 18; translation, 13.
67. B. Koltovoi, in *Nauka i zhurnalist: sbornik*, Vol. 2 (Moscow: Znanie, 1976), 35.
68. A. Averianov and Z. Orudzhev, "Nauka i zhurnalizm," *Zhurnalist*, 10 (1982), 57.
69. V. Druin, in *Nauka i zhurnalist*, 79.
70. "Discussion of the Magazine *Voprosy filosofii* in the USSR Academy of Sciences' Institute of Philosophy," *Vf*, 6 (1974), 157–69; translation, *CDSP*, 45 (1984), 9 (emphasis added).
71. V. Kiriasov, "Take a Good Look at Your Area," *Pravda*, March 23, 1978, 2; translation, *CDSP*, 12 (1978), 24.
72. Mickiewicz, *Media and the Russian Public*, 6.
73. "60 Minutes," CBS News transcript, July 29, 1984, 12.
74. Zhores Medvedev, *Soviet Science*, 193 (emphasis in original).
75. Joshua Rubenstein, *Soviet Dissidents: Their Struggle for Human Rights* (Boston: Beacon Press, 1980), 116–17.
76. Zhores Medvedev, in Zhores and Roy Medvedev, *A Question of Madness* (New York: Knopf, 1971), 175 (emphasis added).
77. Roy Medvedev, in *ibid.*, 44.
78. Andrei Sakharov et al., "An Open Letter to the Central Committee of the CPSU," *Survey*, 76 (Summer, 1970), 164.
79. *Ibid.*, 163; and Andrei Sakharov, *Progress, Coexistence, and Intellectual Freedom* (New York: Norton, 1979), 62–63.
80. Roy Medvedev, *On Socialist Democracy* (New York: Norton, 1975), 164.
81. *Ibid.*, 123, 206 (emphasis added).
82. Ia. Kashlev, "International Relations and Information," *IA*, 8 (1978), 82; and Ia. Zakharov, "International Cooperation and the Battle of Ideas," *IA*, 1 (1976), 86.
83. See, e.g., Zakharov, *op. cit.*, 86.
84. V. Korobeinikov, "What Is Behind the 'Freedom of Information' Concept?" *IA*, 2 (1976), 106.
85. See, e.g., Zakharov, *op. cit.*, 88.
86. See, e.g., G. Kh. Shakhnazarov, "New Factors in Politics at the Present Stage," 48–49.
87. See, e.g., Shakhnazarov, "New Factors in Politics at the Present Stage," 49.
88. Roy Medvedev, *On Socialist Democracy*, 208.
89. "Conference on Security and Cooperation in Europe: Final Act, Helsinki, 1975," *The Department of State: Bulletin Reprint* (Washington, D.C.: 1975), 16–18 (emphasis in original).
90. Robert Sharlet, "Dissent and Repression in the Soviet Union," *Current History*

(October 1977), 130.
91. V. Chkhikvadze, "Human Rights and Noninterference in the Internal Affairs of States," *IA*, 12 (1978), 23.
92. Medvedev, *On Socialist Democracy*, 203.
93. S. N. Smirnov, unattributed quote in *ibid.*, 385–86.
94. Stephen F. Cohen, in Cohen, ed., *An End to Silence: Uncensored Opinion in the Soviet Union* (New York: Norton, 1982), 8.
95. Sharlet, ed., *The New Soviet Constitution of 1977*, 93.
96. N. V. Krylenko, *Rechte und Pflichten des Sowjetburgers* (Moscow: 1936), 31, quoted in George Kline, "Comment," in Peter Potichnyj, ed., *Papers and Proceedings of the McMaster Conference on Dissent in the Soviet Union* (Hamilton, Ontario: 1972), 116.
97. L. F. Ilichev, "Tvorit' dlia naroda, vo imia kommunizma," *Pravda*, December 22, 1962, 2.
98. See, e.g., Alexander Shtromas, *Political Change and Social Development: The Case of the Soviet Union* (Frankfort: Verlag Peter Lang, 1981).
99. Roy Medvedev, in Zhores and Roy Medvedev, *A Question of Madness*, 205–6.
100. *Ibid.*, 207–8 ff.
101. See, e.g., Stephen Weiner, "Socialist Legality on Trial," in Abraham Brumberg, ed., *In Quest of Justice* (New York: Praeger, 1970), 39–51 (emphasis added).
102. Valentyn Moroz, "A Report from the Beria Reservation," in Michael Browne, ed., *Ferment in the Ukraine* (New York: Praeger, 1971), 124.
103. Roy Medvedev, *Let History Judge* (New York: Knopf, 1972), xxvi, xxxi ff.
104. Andrei Amal'rik, *Will the Soviet Union Survive until 1984?* (New York: Harper and Row, 1971).
105. Sakharov, *Progress, Coexistence, and Intellectual Freedom*, 27.
106. Andrei Sakharov, "Memorandum," *Survey*, 3 (Summer 1972), 226.
107. *The New York Times*, March 3, 1974, 1; April 15, 1974, 1.
108. Aleksandr Solzhenitsyn, "Letter to the Fourth Congress of Soviet Writers," in Brumberg, ed., *op. cit.*, 247.
109. Aleksandr Solzhenitsyn, *The First Circle* (New York: Harper and Row, 1968), 358 (emphasis added).
110. See note 107.
111. Aleksandr Solzhenitsyn, *Gulag Archipelago*, 3 vols. (New York: Harper and Row, 1975).
112. Valentyn Moroz, in Browne, ed., *op. cit.*, 134, 138–40 (emphasis in original).

Conclusion: Technocratic Socialism

1. See Hoffmann and Laird, *The Politics of Economic Modernization in the Soviet Union*, and *idem*, *"The Scientific-Technological Revolution" and Soviet Foreign Policy.*
2. This term is Jeffrey Straussman's. See his *The Limits of Technocratic Politics* (New Brunswick, N.J.: Transaction Books, 1978), 23–38 ff.
3. Jurgen Habermas, *Theory and Practice* (Boston, Beacon Press, 1974), 282.
4. Shakhnazarov, *Sotsialisticheskaia demokratiia*, 53, 102; translation, 35, 64.
5. *Ibid.*, 110 ff.; translation, 69 ff. Also, Brezhnev, in *Materialy XXIV s'ezda KPSS*, 57.
6. Shakhnazarov, *Sotsialisticheskaia demokratiia*, 111–12, 124, 178; translation, 69–70, 77, 108 (emphasis in original).

7. *Ibid.*, 70; translation, 45 (emphasis added).
8. *Ibid.*, 121; translation, 75.
9. Roger Garaudy, "Freedom and Creativity: Marxists and Christians," *The Teilhard Review*, 2 (Winter 1968/1969), 46.
10. Roger Garaudy, *Esthetique et Invention du Future* (Paris: Union Generale D'Editions, 1968), 8, 10.
11. *Ibid.*, 12, 23 ff.; and Roger Garaudy, *L'Alternative* (Paris: Robert Laffont, 1972), 153–54 ff.
12. Roger Garaudy, *The Crisis in Communism* (New York: Grove Press, 1970), 29 ff.
13. Garaudy, *Esthetique et Invention du Future*, 9.
14. Garaudy, *L'Alternative*, 88.
15. For a Soviet critique of Garaudy, see, e.g., Kh. N. Momdzhian, *Marxism and the Renegade Garaudy* (Moscow: Progress, 1974).
16. Radovan Richta, *Civilization at the Crossroads*, 253–54, 256 (emphasis in original).
17. Roy Medvedev, *On Socialist Democracy*, p. 231.
18. *Ibid.*, 300.
19. *Ibid.*, 172, 194.
20. *Ibid.*, 122.
21. *Ibid.*, 40, 46–47, 123–24.
22. Brief earlier comparisons of Soviet and Western theories appeared in Erik P. Hoffmann, "Contemporary Soviet Theories of Scientific, Technological, and Social Change," *Social Studies of Science*, 9 (1979), 108–10; Laird, "'Developed' Socialist Society and the Dialectics of Development and Legitimation in the Soviet Union," 131–32; and Hoffmann and Laird, *The Politics of Economic Modernization in the Soviet Union*, 53–56.
23. For Soviet assessments of Bell, see, e.g., G. I. Ikonnikova, *Teoriia "postindustrial'nogo obshchestva": buduschee chelovechestva i ego burzhuaznye tolkovateli* (Moscow: Mysl', 1975); Iu. K. Ostrovitianov, "Postindustrial'naia tsivilizatsiia ili kapitalizm v 2000 godu?" *Vf*, 7 (1969), 30–41; Iu. A. Krasin, "Apologeticheskaia sushchnost' teorii postindustrial'nogo obschchestva," *Vf*, 2 (1974), 55–67.
24. For an interpretation of Soviet views on the New Left, see Klaus Mehnert, *Moscow and the New Left* (Berkeley: University of California Press, 1975).
25. See, e.g., Karl Deutsch, "On the Learning Capacity of Large Political Systems," in Manfred Kochen, ed., *Information for Action: From Knowledge to Wisdom* (New York: Academic Press, 1975), 61–83; Deutsch, *The Nerves of Government*, 2d ed. (New York: The Free Press, 1966).
26. Daniel Bell, *The Cultural Contradictions of Capitalism* (New York: Basic Books, 1976), xi–xii ff.
27. E. A. Arab-Ogly, *V labirinte prorochestv* (Moscow: Politizdat, 1974), 65; translation, *In the Forecaster's Maze* (Moscow: Progress, 1975), 72.
28. Bell, *The Coming of Post-Industrial Society*, xi–xii (emphasis in original).
29. J. K. Galbraith, *The New Industrial State*, 3d ed. (Boston: Houghton Mifflin, 1978), 405, 411. See also Galbraith's *Economics and the Public Purpose* (Boston: Houghton Mifflin, 1973).
30. Shakhnazarov, *Fiasko futurologii*, 66–68; translation, 44–45.
31. Bell, *The Coming of Post-Industrial Society*, xxi (emphasis in original).
32. Cf. Afanas'ev's writings with Deutsch, "On the Learning Capacity of Large Political

Systems"; and Deutsch, *The Nerves of Government.*

33. Yehezkel Dror, *Public Policy Reexamined* (New Brunswick, N.J.: Transaction Books, 1982).
34. Charles Lindblom, *The Intelligence of Democracy* (New York: The Free Press, 1965); David Braybrooke and Charles Lindblom, *A Strategy for Decision* (New York: The Free Press, 1963); and Charles Lindblom, *Politics and Markets: The World's Political-Economic Systems* (New York: Basic Books, 1977).
35. Bell, *The Coming of Post-Industrial Society*, 365.
36. See, e.g., Galbraith, *The New Industrial Society*; and Bell, *The Coming of Post-Industrial Society.*
37. Langdon Winner, *Autonomous Technology: Technics-out-of-Control as a Theme in Political Thought* (Cambridge, Mass.: MIT Press, 1977), 226, 229, 238 ff. (emphasis in original). The term "reverse adaptation" is Winner's.
38. I. T. Frolov, "Perspektivy cheloveka," *Vf*, (1975) 83–95; and *Vf*, 8 (1975), 127–38.
39. G. N. Volkov, *Man and the Challenge of Technology*, 214.
40. Cf. Jean-Jacques Salomon, *Science and Politics* (Cambridge, Mass.: MIT Press, 1973), 177 ff.
41. I. Gerasimov, *Man, Society, and the Environment* (Moscow: Progress, 1975), 24.
42. L. P. Lysenko, *Priroda i obshchestvo* (Minsk: Izdatel'stvo BGU, 1976), 110.
43. See, e.g., A. D. Ursul, *Chelovechestvo, zemlia, vselennaia* (Moscow: Mysl', 1977).
44. S. I. Popov, *Socialism and Humanism* (Moscow: Novosti, 1977), 261 (emphasis added).
45. Fedoseev, "Social Significance of the Scientific and Technological Revolution," 101 (emphasis added).
46. G. N. Volkov, *Man and the Challenge of Technology*, 241.
47. Afanas'ev, *Nauchnoe upravlenie obshchestvom*, 2d ed., 112.
48. V. G. Afanas'ev, *Scientific Communism* (Moscow: Progress, 1967), 195 (emphasis added).
49. Afanas'ev, *Nauchnoe upravlenie obshchestvom*, 2d ed., 295.
50. *Ibid.*, 307–9 ff.
51. *Ibid.*, 235.
52. Sakharov, *Progress, Peaceful Coexistence, and Intellectual Freedom*, 60.
53. Konstantin Simis, *USSR: The Corrupt Society* (New York: Simon and Schuster, 1982), 24, 26, 34.
54. For elaboration of this argument, see W. Brus, *Socialist Ownership and Political Systems* (London: Routledge and Kegan Paul, 1975).
55. Cf. especially Chapter 1, "Theory and Practice," 18–27.
56. This term is John Joyce's. See his "The Old Russian Legacy," *Foreign Policy*, 55 (Summer 1984), 133 ff.
57. *Ibid.*
58. Stephen F. Cohen, in Cohen, Rabinowitch, Sharlet, eds., *op. cit.*, 25.
59. Alvin Gouldner, *The Dialectic of Ideology and Technology: The Origins, Grammar, and Future of Ideology* (New York: Seabury Press, 1976), 289–93 (emphasis in original).
60. G. N. Volkov, *Man and the Challenge of Technology*, 215.
61. Gouldner, *The Dialectic of Ideology and Technology*, 241. (emphasis in original).
62. Bailes, *Technology and Society under Lenin and Stalin*, 408.
63. Emmanuel Mesthene, "The Study of Technology and Society: Methods and Issues," in

Harvard University Program on Technology, 1964–1972: A Final Review, (Cambridge, Mass.: Harvard University Press, 1972), 9. For a critique of Mesthene and the Harvard project, as well as observations pertinent to Soviet theory, see William Leiss, "Technology and Instrumental Rationality in Capitalism and Socialism," in Fleron, ed., *op. cit.*, 115–45; and Leiss, "The Social Consequences of Technological Progress."

Index

About the Authors

Erik P. Hoffmann is Professor of Political Science, The Nelson A. Rockefeller College of Public Affairs and Policy, State University of New York at Albany, and Senior Associate, Research Institute on International Change, Columbia University. He is coauthor of *"The Scientific-Technological Revolution" and Soviet Foreign Policy* and *The Politics of Economic Modernization in the Soviet Union*; editor of *The Soviet Union in the 1980s*; and coeditor of *Soviet Foreign Policy in a Changing World*, *The Soviet Polity in the Modern Era*, and *The Conduct of Soviet Foreign Policy*.

Robbin F. Laird completed his research for this book while a fellow at the Research Institute on International Change, Columbia University. He has recently become affiliated with the Institute for Defense Analyses. He is author of *France, The Soviet Union, and the Nuclear Weapons Issue*; coauthor of *The Soviet Union and Strategic Arms*, *"The Scientific-Technological Revolution" and Soviet Foreign Policy*, and *The Politics of Economic Modernization in the Soviet Union*; and coeditor of *Soviet Foreign Policy in a Changing World* and *The Soviet Polity in the Modern Era*.